Divine Golden Ingenious

The Golden Ratio
as a Theory of Everything?

Divine Golden Ingenious

The Golden Ratio as a Theory of Everything?

HIRMER

Content

Workshop Reports

Divine Golden Ingenious
The Golden Ratio as a Theory of Everything?

Lieselotte Kugler

During my time studying history and art history I frequently came across the golden ratio, and have continued to do so in the course of over thirty years of curatorial work in four different cultural history museums. And just as frequently, I have found myself regarding this phenomenon with both fascination and scepticism. The golden ratio is intimately entangled in the questions of whether a universal formula could exist for ›beauty‹, and whether it could be taken as a valid standard or criterion for all proportions. Moreover, could this formula, the ›ingenious‹ proportion of the golden ratio, ultimately prove to be an essential element of beauty? If that were so, anything designed by applying the golden ratio would count as beautiful, and people could use this criterion, found all around them, as a means of training their eye. They could then clearly say whether the proportions of an object, an art work, a piece of architecture, a flower or an animal were objectively ›beautiful‹ or not. That ›school of seeing‹ found in the world's museums would then take on a very different meaning. Not only in daily life or in architecture, but especially in exhibitions, we would only encounter categories of artefacts and objects judged to be either beautiful or not.

But wouldn't life be boring and uninspiring if everything followed a single standardised formula? Where would be those surprising experiences of beauty, where would be the dramatic moments and contrasts, for example, when looking at art works? Should this harmonious ratio be the measure of all things?

The debate and discussion on the golden ratio also ›divides‹ academia, art and design into two camps – with advocates fiercely arguing for the existence and relevance of a golden ratio for the phenomenon of beauty, and equally vehement opponents arguing against it. Hence, on the one hand, there seems to be no absolutely defined beauty, since there have been and still are various competing theories of proportions. On the other hand, particular phenomena do exist which appear to follow the golden ratio and the Fibonacci sequence, and mathematical proofs provide plausible constructs of the ›divine proportions‹ they contain. Quite aside from all these diverse schools of thought, today's scholarship has incontrovertibly shown that the perception of beauty across the historical eras has undergone various changes, and that a rational and empirical standard of measurability is hardly feasible. So has the correct formula of aesthetics, the code, simply not yet been found? Or does beauty really lie in the eye of the beholder?

In fact, down the centuries there has been lively scholarly interest in proving the existence of the golden ratio in certain natural phenomenon, the hard sciences and art. From the time the golden ratio was first elevated to a ›theory of everything‹ in the nineteenth century, theoretical tracts by artists, philosophers, mathematicians, biologists and musicians – at times, with a tendency to mystification or even with an esoteric touch – have been permeated by a belief in the existence of a corresponding natural law, one providing the basis for the golden ratio, which itself could be found in aesthetic forms in many areas of life. The interdisciplinary nature of the search for the meaning of the golden ratio, both historically and today, is fascinating. For this reason, it was my long-cherished wish to stage an exhibition on this topic, and to cover a broad spectrum of the related issues and connect them to communication. There can be little doubt that few people have addressed the mathematical relations of the golden ratio or dealt with its concrete applications as thoroughly as in this exhibition – which is why it can rightly be expected to appeal to a wide audience.

Quite surprisingly, looking back, there has been no major cultural historical exhibition with an interdisciplinary approach to the golden ratio. In that sense, it seemed rather adventurous to take this as an exhibition topic, especially in the context of ›communication‹ and with the aim of answering such questions as: Why are some designs more successful than others? Why do these products appeal, while those say nothing? Why were – and are – functional objects designed in one way, and not in another? Why do we find particular colour combinations and proportions pleasing, and not others? These and many other questions relating to the communication between subject and object are worth investigating and worthy of being shown in an exhibition. It is certainly no coincidence that the credo of ›communicating by proportions‹ was the leitmotif during the intensive period of preparation for this exhibition.

Since the golden ratio was first mentioned in Greek antiquity in Euclid's *Elements*, it has had an enduring fascination. Aside from the question of beauty and harmony, the main focus has been on the search to identify a universal constant – that the ›theory of everything‹ which the golden ratio seemed to indicate must exist. The exhibition is especially dedicated to the wonder of this ›search‹. It localises its origins, offers a first, fascinating look at imaginatively orchestrated displays, traces the development of mathematical formula and, at the same time, explains traditional misinterpretations. But the exhibition also considers the significance that the golden ratio still has today, offering detailed insights into the role this principle plays in modern design processes and asking how far the golden ratio is relevant for visual communication strategies.

One key section of the exhibition focuses on the scholarly and artistic preoccupation with the golden ratio in the nineteenth century. In those years, as already mentioned, the golden ratio became widely known and discussed, with Gustav Fechner even developing an empirical psychological study designed to confirm the golden ratio's uniqueness and outstanding aesthetic appeal. The exhibition's ›showcase‹ clearly elucidates the reception of the golden ratio today, and shows how it serves as a starting point for various design tasks in architecture, graphics, product design and fashion. In this way, the exhibition traces the arc of a dynamic development from the classical world to today – with the outcome still open.

When we approached various scholars and scientists, artists and graphic artists, designers, biologists, mathematicians and architects with our idea for this exhibition, we were met with such strong interdisciplinary interest that the plan was soon born for an accompanying volume of essays. These are intended to go beyond the actual exhibition itself to provide a comprehensive and lasting overview of the many applications of the golden ratio, and the theories associated with it. The most varied positions are presented here in a wide range of formats – from essays by acknowledged experts ›On Seeking and Finding the Golden Ratio‹ to the ›Workshop Reports‹ analysing specific objects and design and research processes. In addition, a series of ›Formative Proportions‹ present various forms of standardisation, from paper to *tatami* mats, while the ›5 Theses‹ as well as ›In Conversation‹ offer an interesting range of views on the golden ratio. In keeping with this diversity of disciplines and contributors, the approaches to the much-disputed golden ratio are also very different, from scientific meticulousness to an emotional plea for the power of the designed environment on its beholders.

At this point, I would like to express my profound thanks to all those involved in making this remarkable exhibition a reality, from the contributors to the catalogue to the entire museum team. In their very different ways, they have all played their part in making this project a success. In addition, I would especially like to thank the staff and students of the BEST Sabel School of Design Berlin who, as the result of a fruitful cooperation, have contributed valuable insights into the real-world application of the golden ratio, to Brousse & Ruddigkeit Berlin, who have given the catalogue its distinctive design, the Hirmer Publishers Munich for the publication of this accompanying volume and, last but no means least, Franke | Steinert Berlin who accompanied the long journey of this project from planning to reality and expertly designed the exhibition.

Divine Golden Ingenious – The Golden Ratio as a Theory of Everything? is a temporary exhibition. It will also be shown at other venues including our other Museums for Communication in Nuremberg and Frankfurt.

From Aurea to Zeising

In search of the golden ratio

Oliver Götze, Katharina Schillinger

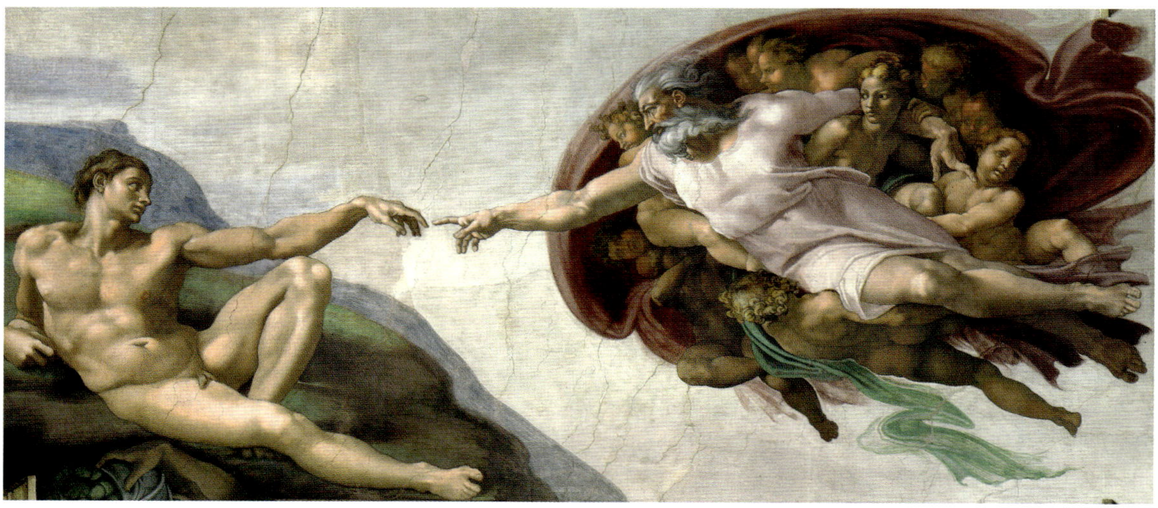

Figure 1. ***The Creation of Adam*****,**
Michelangelo, c. 1512

During the *Am I Beautiful?* exhibition in the winter months of 2013/14, visitors to the Museum for Communication Berlin could vote for the one rectangle out of five that they thought was most harmonious. Here, the Berne museum team designing the exhibition were inspired by a series of experiments by Gustav Theodor Fechner, the German pioneer of perceptual psychology. In 1876, Fechner asked a total of 390 test subjects to select the shape they found most pleasing from ten rectangles of the same surface area. The findings showed a distinct preference for the ›golden rectangle‹ – the rectangle whose sides are exactly in the golden ratio of 1:1.618. There may have been fewer rectangles to choose from in the *Am I Beautiful?* exhibition, but the result was the same. From 13,315 votes counted in total, 3,997 visitors (30 %) voted for the rectangle in the golden ratio, and 3,678 (27.6 %) for the square in the standardised DIN format.

Naturally, the conditions for this experiment solely designed for an exhibition do not meet accepted standards in the empirical social sciences. The exhibition visitors could have voted several times or may have not voted at all, there was no control group, and no information about the visitors' prior knowledge. Nonetheless, the survey was remarkable simply for the size of the random sample, and it produced a result which seems to confirm Fechner's original findings. So in spite of the criticism of the experimental setting, does this prove a preference for the golden ratio among the exhibition visitors – a preference repeatedly posited and claimed to be biologically determined and culturally inherited? Or conversely are industrial standards having such a strong influence that they will soon displace traditional modes of visual perception? Is it possible to develop an exhibition between these two antithetical views, starting from the allegedly most attractive of all proportions to consider a shift in visual preferences?

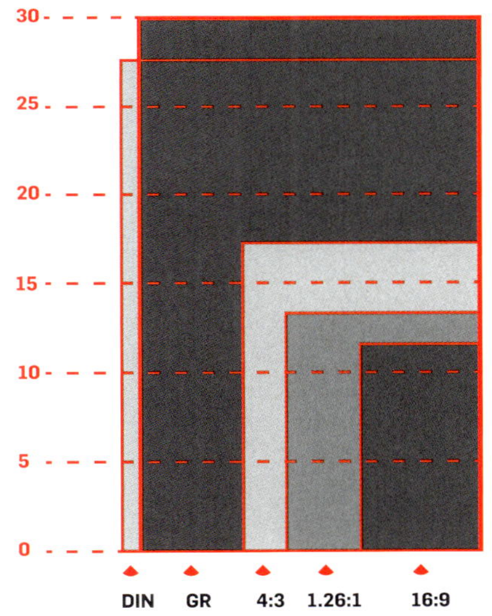

On closer examination, though, the interpretation of the experiment is not that simple. The survey findings cannot be generalised to support a continuing preference for the golden rectangle. Moreover, scientifically conducted replications of Fechner's experiment have obtained different results (e.g. a preference for the square, see Höge 2008). In addition, the status of the golden ratio as the most excellent and pleasing proportion is highly controversial. And finally, anyone who delves into the history of the golden ratio and its use will soon discover that virtually no other subject in the fine arts has produced such staunch proponents and equally intractable opponents. Evidently the golden ratio can only either be enthusiastically supported as a universally valid principle of beauty adopted from the natural world which, from time immemorial, has influenced the design process and still does so today – or be just as vehemently rejected as part of the pseudo-scientific raiment which modern mystics love to dress the world in. The protagonists in this debate appear unable to reach a consensus, and so any exhibition project on this topic also has to be located in this contested space.

An ancient phenomenon?

How traditional is the golden ratio? The first step facing those organising an exhibition on the golden ratio is to decide where they stand on its history. According to the current reading of art historical sources, the golden ratio is evidently far younger than the name and the range of evidence quoted in the specialist literature would suggest. Etymologically, the golden ratio can be traced to a confusion caused by the mathematician Martin Ohm in the second edition of his *Elementarmathematik* (1835) when he mistakenly assigned the ›gold attribute‹ of the *regula aurea* (rule of three) to the continuous division known as the *sectio divina*. Since this mistake was soon also incorporated into various textbooks in mathematics, the golden ratio had become common knowledge in elementary maths as early as around 1860. The idealisation of the proportional aesthetics of the golden ratio at almost the same time was driven by nineteenth-century amateur researchers who, in the spirit of idealist philosophy, turned to various dimensional ratios in their hope of discovering a theory of everything. The work of Adolph Zeising, in particular, plays a key role in the history of the golden ratio. In his *Neue Lehre von den Proportionen des menschlichen Körpers* (New Theory of Proportions of the Human Body), published in 1854, he was the first to comprehensively measure ancient edifices, Renaissance paintings, the heavenly constellations and organic bodies with a proportional compass. His extensive survey discovered the desired proportional relationship between prominent points in all these works, though, admittedly, the points he chose were often arbitrarily selected (Fig. 2). Zeising's work was a surprising success, with an impact hard to overestimate. Thanks to that success, the search for the golden ratio developed into a popular pastime which – as the New Year's Eve photo in Manchester (p. 204) shows – is still keenly pursued today. In addition, it turned the question of an objective ideal of beauty, derived from geometry and the natural world, into one of the key issues in the history of European design. This, in turn, lent a particular significance to constructing a tradition reaching as far back in time as possible, helped by the fact, at least, that the mathematical principles underlying the golden ratio can be traced back to Greek antiquity.

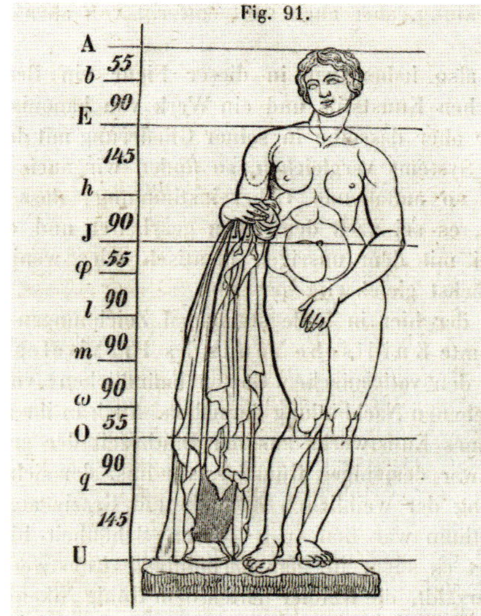

Figure 2. Adolph Zeising's measurement of the *Aphrodite of Knidos by Praxiteles*, 1854

The relation between three quantities as a continuous division was already described in Euclid's *Elements*, and used in geometric proofs. However, in written sources, this was first raised to the status of a ›miraculous‹ (*mirabilis*) proportion in a commentary by Campanus (Elements, XIV, 10, around 1255). Moreover, here too, the golden ratio was praised primarily for its usefulness in constructing regular bodies. There are no known records from the ancient world or the Middle Ages of any aesthetic statements relating the golden ratio to architecture or art. As a result, no matter how convincingly derived, every piece of evidence for the principle's deliberate application as the ideal proportion of beauty – for example, in the display facades of the Parthenon or Gothic cathedrals – is not just speculation, but also quickly finds itself pushed towards the corner of obscure mysticism.

The lack of documented sources has generated laboured arguments of some sort of ›masons' guild secret‹ or an oral tradition which has gradually been lost. Since ultimately, in combination with subjectively identified proportional networks, such arguments can be applied to any arbitrary selection of buildings and art works, the literature on the topic often contains truly bizarre examples claiming to show ›golden‹ ratios. Art historiography may have criticised such efforts ever since Zeising's day, but for a long time such criticisms were quite hesitant. Only since the 1990s have there been more vigorous attempts to counter this myth-making by critically reviewing the sources (Fredel 1993/98) or exposing the arbitrary practices of the ›disciples of the golden ratio‹. Here, for example, George Markowsky's sardonically exaggerated review of misconceptions about the divine proportion (1992) must sow doubts in the mind of anyone searching for the golden ratio. Markowsky shows the impact of lines sketched in bold where possible, details added or omitted, and how with inaccuracies in drawing and measurements the golden ratio can be found in almost every art work. Yet the effect is decidedly modest in the face of primeval beliefs in secret structures and the internet's potential for viral dissemination. Anyone starting to research the golden ratio online – as is common today – has to search extensively among a range of abstruse theories before stumbling across well-founded approaches.

The ›divine proportion‹ of the Renaissance

The golden ratio may first have been extensively investigated in the Renaissance. Not only did the renowned artist and mathematician Piero della Francesca (*Trattato dell'abaco*, 1459, fol. 88v) discuss the *proportione avente il mezzo e doi strem*i, but so did his student Fra Luca Pacioli. In his treatise dedicated to the topic, Pacioli even elevated the golden ratio to the *divina proportione* and asked Leonardo da Vinci to provide impressive woodcut illustrations of regular and semi-perfect polyhedrons.

Beyond these three individuals, knowledge of the golden ratio could reasonably be assumed among their friends who were artists, such as Jacopo de' Barbari, who painted the portrait of Pacioli (p. 104) now on show in Naples, or the Nuremberg artist Albrecht Dürer, who also knew Barbari. Many have succumbed to the evident temptation of trying to identify the first use of the golden ratio as a principle of composition in the works of this circle of artists. For example, the columns of the royal palace in Piero della Francesca's *Flagellation of Christ* are more or less divided in the golden ratio (most recently Longo 2015), which could thus underscore the quite credible interpretation of the panel as a temporal juxtaposition of classical and contemporary ages.

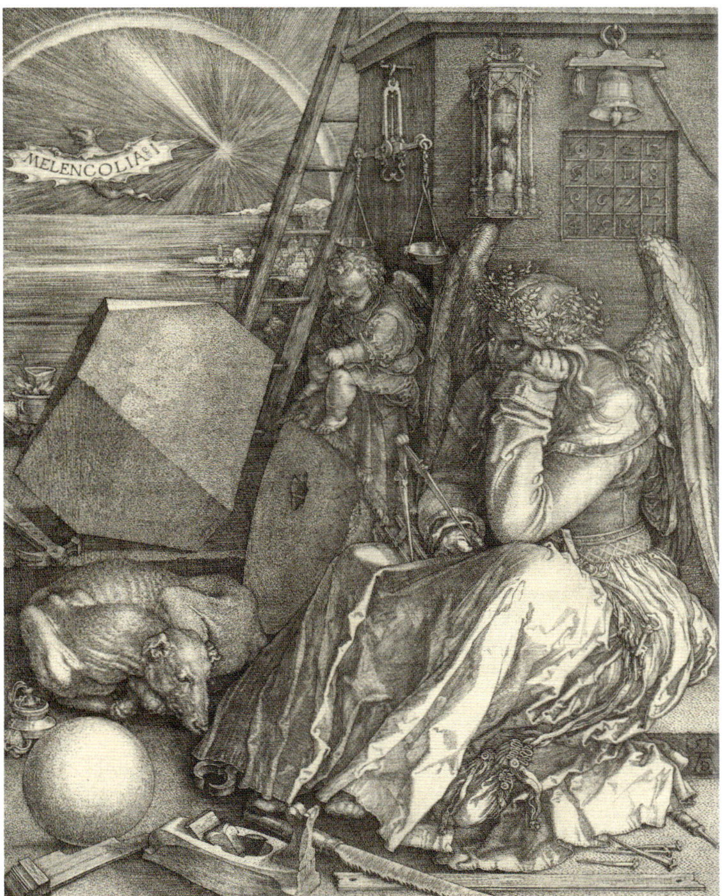

Figure 3. *Melencolia I*, Albrecht Dürer, 1514

Over the last few decades, many examples of golden geometries have been discovered in Leonardo's works, first and foremost in the *Mona Lisa* and the *Last Supper*. Similarly, at the latest since Winziger's study, Albrecht Dürer is also thought to have employed the golden ratio in his *Self-Portrait* (1500) in Munich. The plausibly identified equilateral triangle between the head and shoulders divides the overall height of the painting exactly into the ratio 1.618:1.

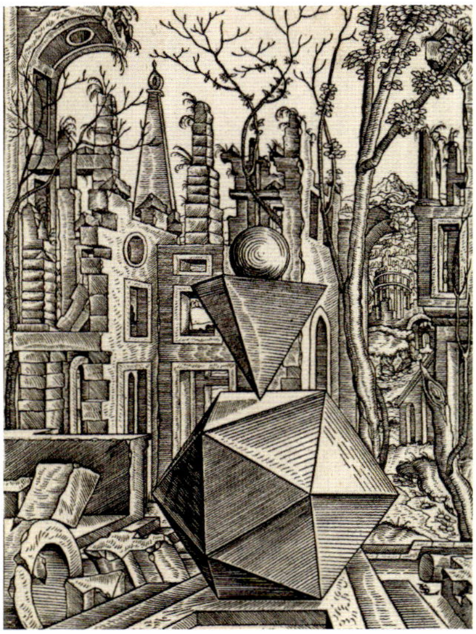

Figure 4. Study of perspective by Lorenz Stoer from *Geometria et Perspectiva*, 1567

In their hunt for the right proportions, though, Zeising's successors did not limit themselves to the artistic circles around Pacioli and Leonardo. Instead, hardly any Renaissance masterpiece seems to have escaped some attempt to prove that the picture was designed around the golden ratio. Michelangelo's ceiling fresco in the Sistine Chapel has only recently joined this illustrious group since the claim is that with the assistance of digital image processing programmes both a latitudinal and transverse divine proportion has been identified between the fingers of God and Adam (Fig. 1, p. 14; De Campos 2015). There is little point in elaborating just how little art historians credit such claims. Yet one should still ask here whether the art historian's painstakingly conducted search for proof – design drawings or written comments by the artists – may sometimes actually obscure the real phenomenon itself. Is it really necessary to show there was a design drawing with the golden ratio for art works or buildings? Or could we not grant great artists an especially developed sense of beauty and harmony intuitively approximating those proportions that the majority of beholders would instinctively regard as beautiful? Might there not even be in Europe some biologically determined preference for placing the defining pictorial subject roughly one fifth away from the centre, as a study at the University of Parma recently attempted to argue (Di Dio 2007)? In the exhibition, visitors can test this on prepared art works, possibly realising for themselves just how open the question of the golden ratio's existence still is, even 150 years after Zeising.

From Pacioli to Kepler

But to return to the history of the golden ratio: Luca Pacioli's *De divina proportione* was the pivotal element of many efforts to plausibly show that the golden ratio had been deliberately used during the Renaissance. Pacioli is far from being an uncontroversial mathematician among historians of science. He was said to be a brilliant speaker, able to hold hundreds of listeners spellbound. At the same time, he was an excellent compiler of knowledge, and knew how to publish successful works skilfully combining ancient and contemporary sources supplemented with illustrations. In contrast, his own mathematical achievements rarely went beyond the ground already laid by Euclid, Vitruvius or Campanus. At first glance, his *divina proportione* also seems exciting, with Pacioli the only one willing to explicitly offer a single solution to the Renaissance ideal of the thought-provoking search for a theory of everything. There is no doubt about his answer – which is the golden ratio – yet when it comes to the details, he fails to live up to his aspirations, only repeating Euclid's proofs and failing to show the golden ratio's application, either practically or aesthetically. This itself suggests that he may not even have been aware of its possible application in art at all. After all, although the second part of his treatise is dedicated to artisan craftwork and architecture, rather than turning to the divine proportion, he discusses Vitruvius' theory of musical harmony. He does mention Leonardo's *Last Supper*, supposedly composed around the golden ratio, but only in the context of the twelve Apostles as a number without any reference to the work's possible proportions. In that sense, an argument for the dissemination of the golden ratio in Renaissance art based on Pacioli's *divina proportione* is more than speculative.

Nevertheless, theories on the use of the golden ratio can provide a convincing impetus to open up new and fascinating readings of art works, as Dürer's famous *Denkbild* (›thought-image‹) *Melencolia I* (Fig. 3, p. 19) exemplarily illustrates. Even in 1970, Siegfried Rösch was suggesting that the curious polyhedron on the left was originally a rhomboid with six faces with the apices truncated so that multiple proportional relations between the sides, diagonals and the radii of the sphere in the same space are exactly in the golden ratio. The question of the sense of such a structure can be answered with the help of art historian Peter-Klaus Schuster's pioneering interpretation of the print. In his view, Dürer has depicted human virtues and the well-founded basis of knowledge on the right-hand side with its tower, scales, hourglass, and so on, while on the left, the water, comet and the ball associated with the goddess Fortuna symbolise the uncertainty of insight. The polyhedron appears to aptly supplement the ideas enumerated on the left-hand side if it is read as a deliberate allusion to irrational numbers – numbers which Pacioli alone described as divine wisdom, while his contemporaries viewed them as far from ›true‹, seeing in them uncertainties and imponderables.

The interpretation above, though, remains speculative, since the interior angles of Dürer's polyhedron, due to the unresolved perspectival distortion, could in principle lie anywhere between 72 and 80 degrees. This does, however, offer a Janus-faced indication of the reception history of the *divina proportione*. While the contemporary scepticism towards an infinite continuous division would certainly have complicated Dürer's acceptance of Pacioli's ideas, they may have fascinated him as a geometric game. As a general assumption, this could equally be claimed for Pacioli's milieu. As Pacioli himself remarked, he was competing at court with astrologists, physicians and artist-architects for the attention of patrons. And in a direct comparison with Renaissance ›tabloid topics‹ – such as astrological portents and prophesies or accounts of the New World – the courtly world of Urbino and Milan may very soon have tired of complex principles governing the qualities of a geometric proportion.

In contrast, the effect of Leonardo da Vinci's woodcut illustrations was quite remarkable; indeed, one could even claim that with the *divina proportione* Leonardo and Pacioli created the basis for the way in which (semi-)regular geometric bodies were rendered in frescos, marquetry or engravings. The famous inlay work by Fra Giovanni in Rome, Siena and Verona, directly inspired by the woodcut illustrations, just as much as the Ambras collection's board game (c. 1570) or Lorenz Stoer's studies of perspective (Fig. 4, p. 20), eloquently testify to this popular sixteenth-century style.

There is indeed certainly something to be said for the theory that the young Johannes Kepler's famous *Mysterium Cosmographicum* (Fig. 5) was inspired by mathematical and geometric art (Brecher 2011). In this work, Kepler presents a cosmological model deriving the distances between the planetary spheres from the proportions of the Platonic bodies, which may well come closest to the ideal of a theory of everything pursued by Pacioli. Yet even if Kepler later rejected this model, becoming the first to break with the two-thousand-year-old axiom of circular and regular planetary movements, he retained a lifelong belief in a cosmos governed by aesthetic principles. Quite apart from its role in his famous late work *Harmonices mundi libri V*, his belief in the harmony of the world informs most of his letters and shorter writings. So it is hardly surprising that, as a mathematician, he more or less inevitably came across the Euclidian construction of the golden ratio, praising it as a ›precious jewel‹, just as Campanus and Pacioli had done.

However, he went far beyond his predecessors' ideas. Not only was Kepler the first to describe the connection between Fibonacci numbers and the *sectio divina* (1608), but in his reflections on the recursive law of the Fibonacci sequence, he also mentioned the number five as symbolising natural propagation and procreation, citing as an example its frequent occurrence in the arrangements of leaves and blossoms (*Strena Seu de Nive Sexangula* /A New Year's Gift of Hexagonal Snow, 1610/11).

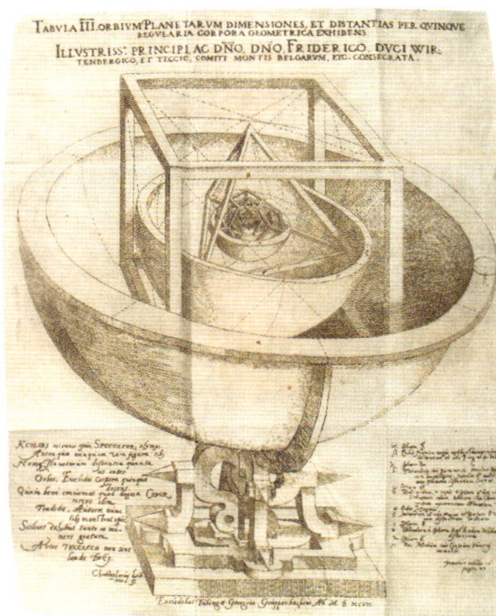

Figure 5. *Mysterium Cosmographicum*, Johannes Kepler, 1596

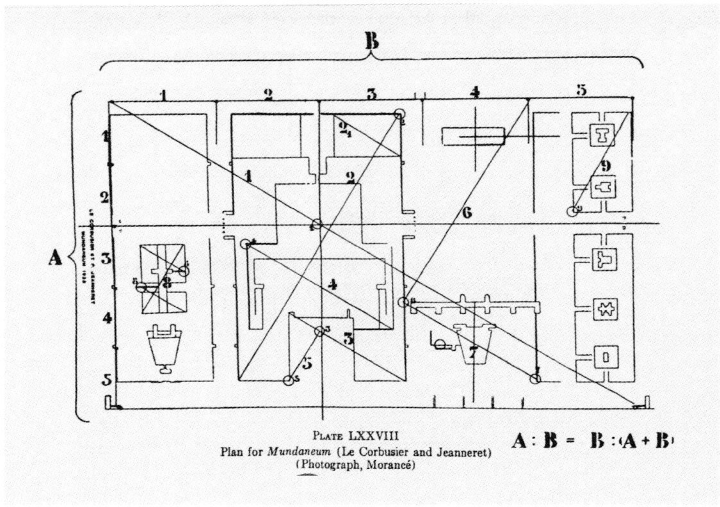

Figure 6. Plan of the Mundaneum marked with Ghyka's proportional lines

Unfortunately, Kepler never further elaborated this initial idea of phyllotaxis in any detail. It was then left to later scientists, such as Bonnet (1754) as well as Schimper and Braun (after 1830), to identify the mathematical laws of leaf arrangements, relating them in some instances to the Fibonacci series – a discovery which may also have influenced Zeising. For a time after Kepler, though, despite the baroque love of geometric shapes and constructions, the fascination for the *sectio divina* appears to have faded again.

In brief, then, taking a sober look at the sources discussed here, the history of the golden ratio prior to the nineteenth century has to be summarised in very modest terms: the golden ratio is a geometric proportional ratio, first described in the classical world, which apparently enjoyed a particular status among mathematicians after the High Middle Ages, but no definite statements can be made about its application in the realm of aesthetics. Admittedly, the significance of the golden ratio today does not lie in real documentary evidence, but in the myth woven around this history over the last 150 years. In common with those searching for the Holy Grail or the treasure of the knights Templar, the golden ratio devotees weave together clues and allusions in history, mathematics and the natural sciences to create grand images which primarily prompt amazement and claim the existence of a universal theory of everything. In the first place, then, any exhibition on the golden ratio has to deal with this myth. For this reason, the *Divine Golden Ingenious – The Golden Ratio as a Theory of Everything?* exhibition starts with the installation of a *Wunderkammer* – a Cabinet of Wonders – containing what are in the meantime seemingly classic objects from the arts, mathematics and the natural world, before the ›Golden Road‹ leads to the turning points in this history.

From myth to application

The suspense between the golden ratio's mystifying idealisation and its practical application is a challenge for any exhibition on this topic. In this context, the question is whether practical applications could not indeed be derived from this influential, constructed myth. Does the much vaunted theory of everything – at least in the post-Zeising era – actually offer an answer to questions of proportion in art, architecture and craft works? The exhibition addresses this contested space by taking visitors past a wall of images presenting the nineteenth-century ›measurement mania‹ to a workshop packed with seemingly promising golden instruments, and then to a display of modern applications.

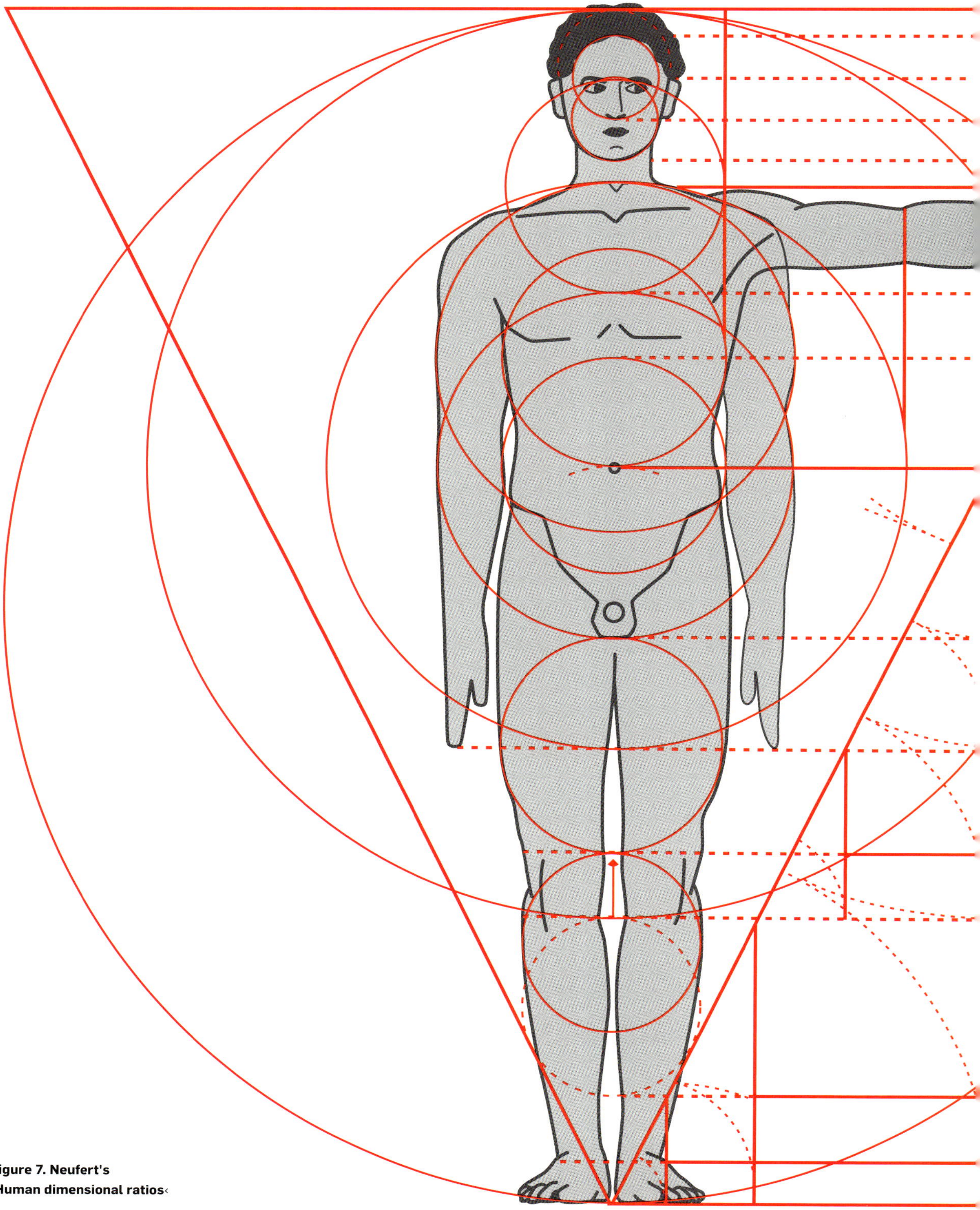

Figure 7. Neufert's ›Human dimensional ratios‹

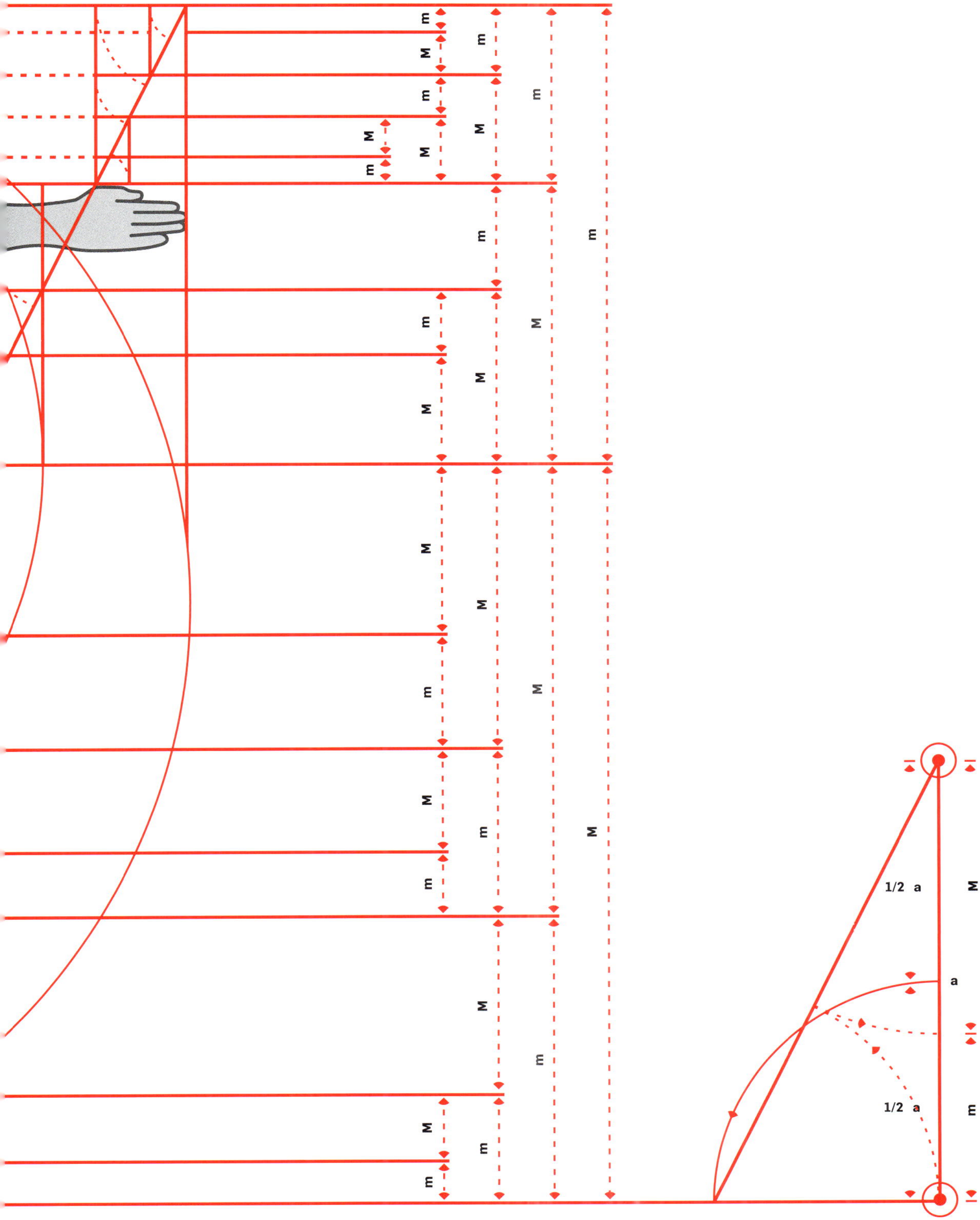
1/2 a
a
1/2 a
M
m

Ultimately, the idea of the golden ratio as a useful tool did become established among Zeising's successors. In 1886, J. Matthias, a teacher at a vocational school, published *Die Regel vom Goldenen Schnitt im Kunstgewerbe* (The Principle of the Golden Ratio in the Arts and Crafts), while in 1919 Rudolf Engel-Hardt derived a set of rules from the golden ratio on how to use it in printing. In 1893, Adalbert Goeringer developed the golden ratio dividers, a calliper with three legs where the central leg divides the distance between the two outer points into the golden ratio. It is, though, almost impossible now to establish just how successful Goeringer's calliper was or how rigorously the rules proposed by Matthias and Engel-Hardt were actually applied. At least we do know that Adolf Hölzel, an artist and contemporary of Goeringer, did use the ›golden dividers‹. Not only did Hölzel write the enthusiastic preface to a publication promoting the dividers, but he also regarded them as an important instrument for schooling the eye during an artist's training. Nonetheless, he never mentioned whether and how he concretely used the dividers in establishing the proportions in his paintings. For the moment, then, the golden ratio remained theoretical, although the idea of implementing this promising theory of everything continued to have its effect.

Division and pattern. The golden ratio in architecture

At first glance, numerical values with innumerable figures after the decimal point seem seriously impractical in any daily application – so this feature of the golden ratio appears almost incompatible with a system of functional architectural patterns. In that sense, the last place one would expect to find references to the golden ratio would be in a book on standardisation and industrial norms.

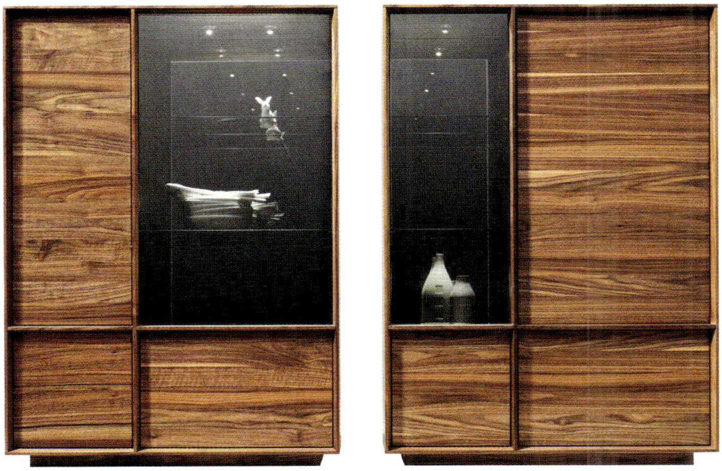

Figure 8. Wall unit *Lux* by Jacob Strobel

Nevertheless, Ernst Neufert's *Bauentwurfslehre* (Architects' Data), first published in 1936 and still a standard reference work for normed initial designs and planning in building projects, not only addressed the golden ratio in the text, but also included it in a very prominent position – as the iconic title page design. In this design, a number of circles are set around and superimposed on a male figure whose right arm is held closely against the body and the left arm horizontally. Initially, it is not clear which system is being used for the distances between the circles. The figure is set within lines suggesting the outline of a square, and with these lines touching the top of the head and bottom of the feet they strongly evoke an association with Leonardo da Vinci's Vitruvian Man. The drawing (Fig. 7, pp. 26/27) is explained in Neufert's book. The auxiliary lines lead to a chain of dimensions with distances marked by M and m.

In other words, the human body is divided into major and minor sections related in the golden ratio, and the origin of that idea is evident from the note »constructed on the proportions given by A. Zeising«. Yet this prominently displayed theory of harmonious proportions does not flow into Neufert's concrete guidelines, and readers can search in vain here for a connection between functional architecture and the golden ratio.

This reference to a traditional theory of harmonious proportions is more like an alibi, legitimising an idea of rationalisation and standardisation, which undoubtedly encountered considerable scepticism at the time. The forty-first German edition of Neufert's *Bauordnungslehre* published in 2016 still contains not only the chapter on the golden ratio, but also the reference to Zeising. The iconic figure has become a brand with a high recognition value, and is often adapted in art and graphic design.

About the same time as Neufert's *Bauordnungslehre*, the Romanian mathematician and philosopher Matila Ghyka published his *Le nombre d'or. Rites et rythmes pythagoriciens dans le development de la civilisation occidentale* (The Golden Number. Pythagorean Rites and Rhythms in the Development of Western Civilisation). His *The Geometry of Art and Life* then appeared around fifteen years later. In terms of their content, Ghyka's writings did not offer any new insights. In the tradition of Zeising, Ghyka measured natural and art objects to provide evidence for a universal theory of everything (Fig. 10, p. 34). There was, though, one new aspect to his measurements – not only did he measure the accepted classics, but also contemporary works including, in particular, a design plan by Le Corbusier for the Mundaneum (Fig. 6, p. 24).

Ghyka's and Neufert's publications virtually symbolise how, in the early twentieth century, the idea of the golden ratio was located in the arc of tension between a mystical theory of everything and practical functionalism. Certainly, both of these authors were influential in encouraging Le Corbusier's intensive investigations into this topic. In the tradition of measuring and re-measuring, and undoubtedly with a longing for »[...] the rule which orders, that connects all things?« (Le Corbusier 1956, p. 26), Le Corbusier started working on geometric constructions from 1918. From 1942, the insights he gained led him to develop his Modulor system of proportion based on a graduated linear scale derived from key values from the human (ideal) frame and the golden ratio. In Le Corbusier's view, the Modulor was capable of solving all design problems, and it indeed seemed to be a successful symbiosis of metaphysical notions of harmony and functional architecture. For the first time in history, the golden ratio became one fundamental element in a comprehensive design concept intended for practical use.

Today, quite a number of architectural offices bear the distinctive Modulor name. The stylised image of the Modulor with the dimensioning system has itself become iconic, taken up in many different ways and re-interpreted. However, as a basis for design, the Modulor is rarely adapted, no doubt due to the complexity of the system – which can confuse and embarrass even experienced architects when asked to explain it. In addition, the Modulor leaves little space for other influencing variables, as Le Corbusier discovered personally in 1957 when he was designing the Berlin *Unité d'Habitation* (p. 67). The planned ceiling heights on the Modulor scale were too low to meet Berlin's building provisions; after months of dispute and discussion, Le Corbusier finally had to give way to the Berlin authorities.

So what is left of the promising idea of a practical design grid? It is certainly true that the golden ratio's presence in the literature of art and, above all, in the theoretical training of designers and architects comes through Le Corbusier, and today is largely connected with his Modulor. Nonetheless, one should also note here that, in an age in thrall to efficiency and rationalisation, Le Corbusier cleverly took up the golden ratio's promise of harmony and tapped into the myth to reinforce the idea behind his Modulor.

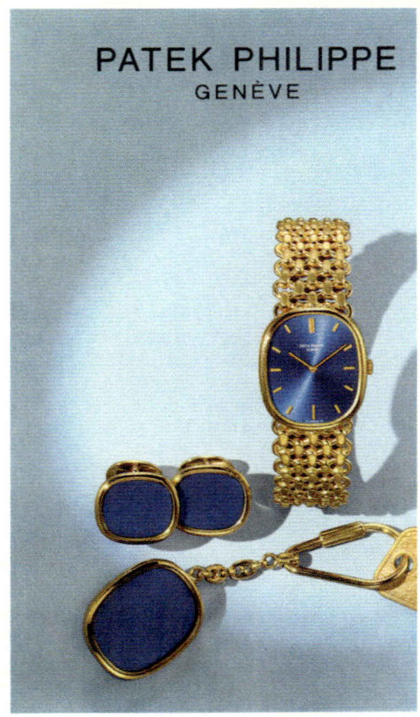

Figure 9a. ***Ellipse d'Or*** **advertising flyer, from 1968**

A guide or a restriction? The golden ratio in product design

Turning to the realm of designed objects, taking a fresh – and rather superficial – look at it without preconceptions, you can find the golden ratio in a surprising number of design classics. For example, Mies van der Rohe's cantilever *Brno* chair fits into a golden rectangle, just as does the rear view of Charles Eames' plywood chair; the Volkswagen *New Beetle* can be almost perfectly set into a golden ellipse, and the *iPod* also fills a golden rectangle. Yet usually questioning these rather arbitrary retrospective measurements and investigating further, searching for evidence of the designers themselves actually using the golden ratio, soon leads to a dead end. Since all subsequent re-measurements without any documented evidence must initially be taken merely as conjecture, just as in Renaissance art or classical architecture, the question arises what this ratio actually means for designers. And to answer that question, it seems reasonable to look at the product design process.

Since the golden ratio can, in fact, be taken as the basis to develop patterns or templates providing points of reference in the infinity of space needing to be defined, you can find real-life examples of its application in manuals and handbooks in this area. However, general overviews for individual design areas only discuss the golden ratio – if at all – as *one* possibility among many. It is rare to find handbooks especially dedicated to the golden ratio and its applications. In 1953, Hellmut Schenck published a slim volume for carpenters and joiners exemplarily illustrating the potential of using the golden ratio, though there is no trace of any pieces made on the basis of his advice.

If we turn to products today, there are certainly examples of designs where one can easily trace the application of such a dimensional pattern. In his *m01* (p. 124) table trestles, Mark Braun has created a strikingly linear piece of furniture which includes proportions in the golden ratio. In its further proportional divisions, the front of Jacob Strobel's *Lux* wall unit (Fig. 8, p. 29) strictly follows a golden rectangle. For more flexibility, one could turn to the Fibonacci series since compared to right-angled surfaces and linear divisions, the Fibonacci series and golden angles facilitate forms whose effect comes closest to natural structures. For example, when looking at Simone Stocker's *Torlo* porcelain set (p. 111), which follows an exactly calculated geometry, the beholder has the vibrant impression of a blossom just opening. Depending on the perspective, the geometry seems to change – which underlines how it is only possible to exactly adhere to a measurement on paper, in two-dimensional space. Three-dimensionality changes the perspective and the product looks different than intended. Nonetheless, a product's attraction may lie exactly in this interplay. The other factors, such as the choice of material, heighten or counter this impression – geometry becomes the experience of form.

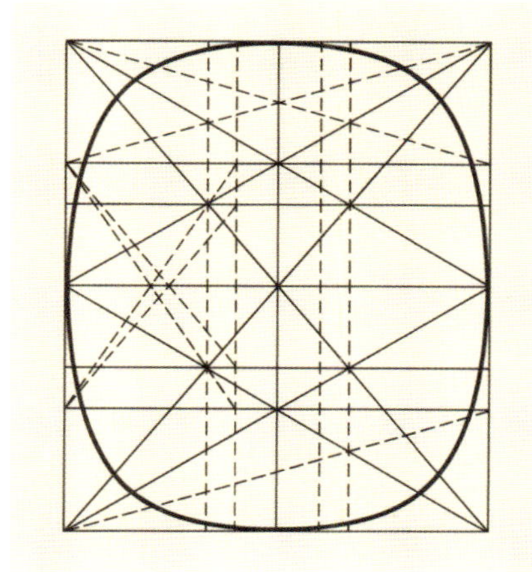

Figure 9b. *Ellipse d'Or* design study, from 1968

This, though, brings out one other aspect: since designed artefacts pass through a very complex development process and are subject to the most diverse demands, it seems more sensible to adopt a rather limited pattern such as the golden ratio as a reference or anchor point than as the basis of a design. After all, geometric and mathematical precision are only some of the many steps on the way from the idea to product development. The crucial factors here range from adhering to industrial standards and regulations to the measurements of the packaging, which also have to be skilfully chosen, just as do the proportions of the product itself. Designs are reshaped, original ideas and patterns disappear, and elements created to meet additional requirements have to be supplemented or simply intuitively aligned.

The search for the golden ratio in one product is, in the first place, always a search for a design process, a strategy, hidden behind the form.

Frequently, it is no longer possible to identify that form in the final product with the naked eye – though that, of course, may be different in works generated more from a fascination with mathematics. The design of Lutz Koebele-Lipp's *Golden Ratio* pendant light (p. 193), for example, grew out of the icosahedron's geometry, and the light's very shape testifies to this fascination with mathematical bodies, evocative of the Renaissance tradition. In this case, the search for a rule for harmony becomes secondary. Rather than the golden ratio being taken as a tool to facilitate composition, here it becomes the element shaping the design itself.

Can the golden ratio or a series based on it ever be more than a dimensioning system offering a point of orientation? From the consumer's perspective, is it ultimately a system which also facilitates evaluation and judgement? Is it genuinely preferred to other methods? In 1997, to find out how product designers themselves would divide a line, Macrosson and Strachan, two researchers in the UK, set up an empirical experiment. According to their findings, the most commonly selected line partitions corresponded to the ratios of 1:1 and 1:2, and there was no evidence for any definite preference. A coherent composition makes it easier for the beholder to grasp the decisions a designer may have taken on the geometry, vocabulary of forms and proportions of a given design. This supports an accessible design process allowing the objects to begin their own narratives – and in the exhibition's showcase visitors can then judge for themselves how convincing these narratives are. Here, visitors can either allow the objects to speak for themselves directly or enlist the help of a golden template. At this point, whether the designer has chosen a particular form consciously or intuitively is not so relevant; what is crucial is that the design of the product does communicate some values.

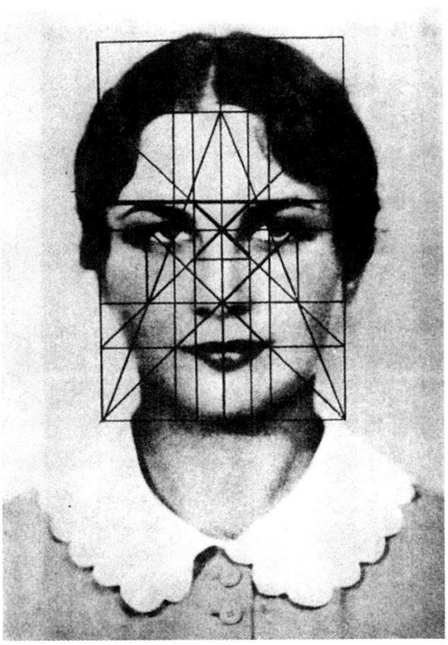

Figure 10. Matila Ghyka ›proves‹ the beauty of the US tennis player Helen Wills Moody

The golden ratio as a message

In a review of the brands advertising at present with the golden ratio – in other words, drawing on this myth –, these are the values which are vaunted: »Beautiful dimensional ratio as formal precept for utilitarian objects« (FSB lever door handle); »Breathtaking proportions – The golden ratio sits at the heart of every Aston Martin«; »Divine Proportions« (Patek Philippe's watch *Ellipse d'Or*, Fig. 9a/9b, pp. 31/33). At times, such claims can be unwittingly funny: »The clear and precise lines of the long and short horizontal pipe legs arranged in the golden ratio offer a harmonious yet exciting appearance« (Aurea heated towel rail).

The advertisements convey the cultural context, a brand's prestige and the quality of the product – so far, so normal. But the golden ratio, deliberately included as a code, gives a promise of exclusivity, a mystical unique selling point. In this way, not only is the product's quality underlined, but it becomes charged with history, tradition and meaning – a constructed tradition which cannot be traced in this form to the ancient world or the Renaissance. This unique selling point plants the idea of specialist knowledge in the mind of the potential group of buyers: »I know a bit about art, culture and history« – higher educational and income levels, and design awareness. Should the history of the golden ratio have played a more secondary role in the design process, this is where it has an even greater (economic) importance – both the term's noble sound and the pseudo-scientific aura it acquired in the nineteenth century create a halo effect around such promises.

Hence, rather than looking for the significance of the golden ratio primarily in the aesthetics of form, it is more likely to be found in the symbolic area of product language and, above all, in the marketing strategies used for advertising products. In addition, when we ask about the intended target group of products advertised in this way, it becomes clear that the products are all located in the luxury sector. The golden ratio is a powerful associative tool, which can be excellently utilised for advertising purposes. This divine proportion may be readily available and straightforward in its use in design, but the products advertised with it are not.

So what is left of the golden ratio in terms of its aesthetic use? In fact, recent history can provide significantly more examples of how the golden ratio has been used as a tool for analysis and training the eye rather than as a concrete design instrument. As Le Corbusier put it, these rules »may be conscious methods, pointed and subtle, or they may be commonplace rules, tritely applied. They may also be [...] a manifestation of [the artist's] intuitive harmony«.[5]

Brecher, Kenneth: »Kepler's Mysterium Cosmographicum. A Bridge Between Art and Astronomy?«, in: Proceedings of Bridges 2011: Mathematics, Music, Art, Architecture, Culture (2011), Phoenix, Tessellations Publishing, pp. 379 ff.

Bürdek, Bernhard E: »Designtheorie«, in: Romero-Tejedor, Felicidad/Jonas, Wolfgang (eds): Positionen zur Designwissenschaft, Kassel: Kassel University Press 2010, pp. 26–31.

De Campos, Deivis et al.: »More Than A Neuroanatomical Representation in The Creation of Adam by Michelangelo Buonarroti. A Representation of the Golden Ratio«, in: Clinical Anatomy, Vol. 28, No. 6 (September 2015), Hoboken, Wiley-Blackwell, pp. 702 ff.

Di Dio, Cinzia/Macaluso, Emiliano/Rizzolatti, Giacomo: »The Golden Beauty. Brain Response to Classical and Renaissance Sculptures«, in: PloS ONE, No. 11, November 2007, 11:e1201, San Francisco, PLOS ONE.

Drach, Ekkehard: Architektur und Geometrie. Zur Historizität formaler Ordnungssysteme, Bielefeld, transcript 2012.

Engel-Hardt, Rudolf: Der goldene Schnitt im Buchgewerbe. Ein Regelwerk für Buchdrucker und Buchgewerbler, Leipzig, Mäser 1919.

Fra Luca Pacioli: Divina Proportione. Die Lehre vom Goldenen Schnitt, edited and translated by Constantin Winterberg, Vienna, Carl Graeser 1889.

Fredel, Jürgen: Maßästhetische Studien zu Proportionsfragen und zum Goldenen Schnitt, Hamburg, Lit 1998.

Ghyka, Matila: The Geometry of Art and Life, New York, Dover 1977.

Höge, Holger: »Keine Schönheit ohne Maß«, in: Zeitschrift für Literaturwissenschaft und Linguistik 152 (2008), Stuttgart, J. B. Metzler, pp. 25 ff.

Karmasin, Helene: Produkte als Botschaften. Konsumenten, Marken und Produktstrategien, Munich, mi [4]2007.

Nonetheless, the search for the golden ratio has not lost its attraction, any more than the search for the *beautiful*. The longing for a formula linking mathematics and art that »holds the world together at its innermost core« (Goethe's *Faust*) is fuelled by a fascination with arcane knowledge and the desire for proof of a ›divine‹ plan in nature – a timeless desire, today just as in the past. The golden ratio can certainly be an instrument, yet always has to be seen in connection and comparison with other methods. It is only one possibility among many. Here, it is always important to retain the clarity of thought to withstand the magic of images, a hagiographical narrative, and a retrospective ascription of meaning.

Ultimately, then, what is primarily interesting may not be the answer to the question of relevance as much as the products to which we address that question – and a cross-section of such products seems a suitable gauge for assessing the mindset of our age. With ever fewer mysteries in our everyday lives, the gap between simplifying our environments and our ever more complex possibilities as well as the anticipated approaching and conclusive de-mystification of the world, the search for the golden ratio seems as anachronistic as it is modern. Visitors to the *Divine Golden Ingenious – The Golden Ratio as a Theory of Everything?* exhibition can explore the questions raised here at the products and exhibits in the ›showcase‹ – and the communication that then takes place will, once again, remain a subjective experience.

Le Corbusier: The Modulor, (reimpression of the first English edition by Faber and Faber (1954), trans. Peter de Francia and Anna Bostock), Basel, Birkhauser, 2004.
Longo, Giuseppe O: »La bellezza sonda della realtà«, in: Pierluigi Barrotta, Scienza e valori: Il bello, il buono, il vero, 2014, Rome, Armando, pp. 125 ff.
Macrosson, W. D. K./Strachan, G.C.: »The Preference Amongst Product Designers for the Golden Section in Line Partitioning«, in: Empirical Studies of the Arts July 1997 vol. 15 no. 2, University of Strathclyde, SAGE 1997, pp. 153–163.
Markowsky, George: »Misconceptions about the Golden Ratio«, in: The College Mathematics Journal, Vol. 23, No. 1 (Jan., 1992), Washington, D.C., Mathematical Association of America, pp. 2 ff.
Matthias, J.: Die Regel vom Goldenen Schnitt im Kunstgewerbe. Ein Handbuch für Werkstatt, Schule und Haus, Leipzig, H. Haessel 1886.
Neufert, Ernst: Bauentwurfslehre. Grundlagen, Normen, Vorschriften über Anlage, Bau, Gestaltung, Raumbedarf, Raumbeziehungen, Maße für Gebäude, Räume, Einrichtungen, Geräte mit dem Menschen als Maß und Ziel, Wiesbaden, Vieweg + Teubner [41]2016.
Rösch, Siegfried: »Die Bedeutung des Polyeders in Dürers Kupferstich ›Melencolia I‹«, in: Fortschritte der Mineralogie 48, Supplement, 1970, Stuttgart, Schweizerbart'sche Verlagsbuchhandlung, pp. 83 ff.
Schröder, Eberhard: Dürer, Kunst und Geometrie. Dürers künstlerisches Schaffen aus der Sicht seiner »Underweysung«, Basel, Birkhäuser 1980.
Schuster, Peter-Klaus: Melencolia I. Dürers Denkbild, Berlin, Gebr. Mann 1991.
Steffen, Dagmar: Design als Produktsprache: Der »Offenbacher Ansatz« in Theorie und Praxis, Frankfurt am Main, form 1999.
Wolf, Brigitte: »Design und die Lehre vom Menschen«, in: Romero-Tejedor, Felicidad/Jonas, Wolfgang (eds): Positionen zur Designwissenschaft, Kassel, Kassel University Press 2010, pp. 207–211.

More than fibre
Paper Formats

Florian Schütz

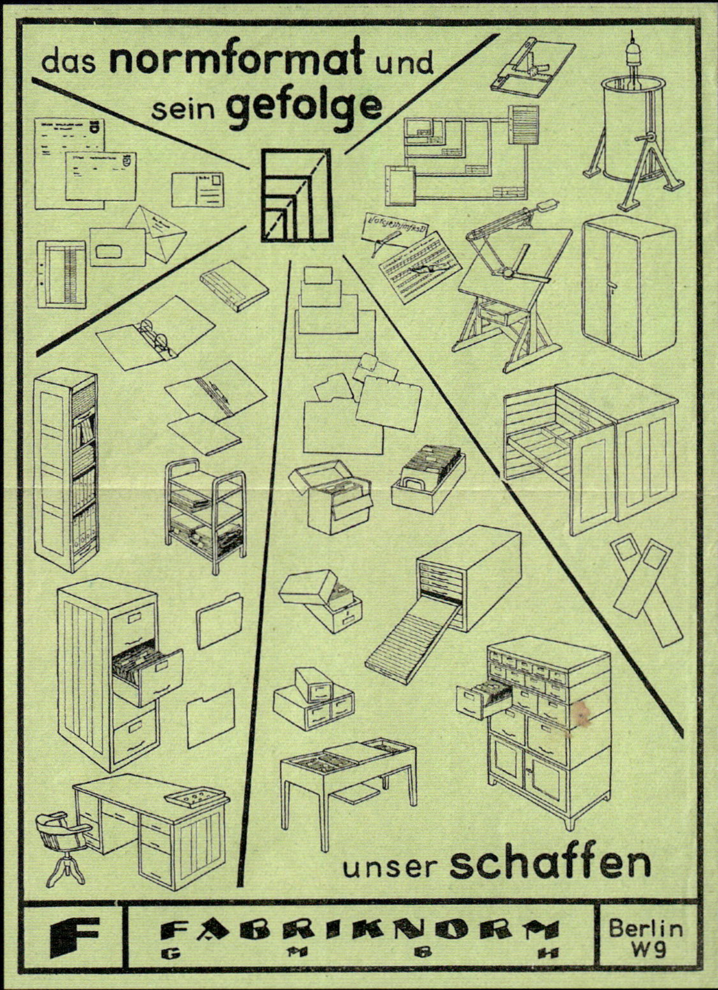

***das normformat und sein gefolge*, Fabriknorm GmbH advertisement, 1920s**

›Paper‹ is a word with many meanings, from absorbent paper to papers for rolling cigarettes, or even identity papers. But usually when we talk of paper, we mean the paper we write on. And if one talks of a ›sheet of paper‹, this conjures up for the majority of people in Europe an image of paper with an A4 format. Or, of course, A3, A2, A6, or A0. But initially at least, it makes no difference which standardised A paper format you think of – and there is a good reason for that.

But first the bureaucratic background. In 1922, Germany introduced the standard DIN 476 as the valid paper sheet format for »paper and paper products for processing data, for offices and schools« – in other words, for all writing and printing paper. The German norm was then taken as the basis for the international (European) standard EN ISO 216, which follows the DIN standards with, in some cases, marginal deviations. In those few countries not using the metric system, such as the United States, paper formats usually take inches as a reference point. As a result, since ISO or DIN norms are calculated in millimetres, these countries apply other standards, with most non-metric countries using the ›letter‹ paper size.

And why does it make no difference which standardised A paper format from A0 to A10 you have in mind? The answer lies in the geometrical relationship between the formats. In this system of standardised sizes, when a sheet of one format is cut in half, it produces two sheets of the next smaller format. The height-to-width ratio of the new pages is then exactly same as the original sheet, a ratio which can be expressed as $1:\sqrt{2}$. The formats produced in this way are geometrically similar, and theoretically the process could be continued infinitely. Even the basis weight or grammage, as it is known, of a standard DIN sheet of paper can be exactly defined. A sheet of A0 with an area of one square metre has to be folded sixteen times to produce one A4 sheet. Hence, to find out the grammage of the A4 sheet, i.e. the paper quality in g/m, you just need to weigh 16 sheets.

In Germany, the leading figure in developing the principles of standardised paper formats was Walter Porstmann. In the early twentieth century, he also wrote on the theory of standardised measurement systems and their efficiency in daily use.

After all, once a product is normed, all products that relate to it have to be congruent with that standard. In 1923, as a result of his work, Porstmann was appointed as managing director of the Berlin company Fabriknorm GmbH. The company manufactured all the related paper products for DIN paper formats, providing the entire office equipment, from filing cabinets to files and envelopes, needed to match those formats.

In 1911, a research institute called *Die Brücke*, which was dedicated to organising intellectual work, had already produced a *Weltformat* (World Standard). This also applied the same 1:√2 principle as did later DIN formats, but never became widely accepted. Since the letter format deviated marginally, it would not fit the file sizes common in that day. Nonetheless, the poster size in Switzerland still follows the *Weltformat* standards.

The ›paperless office‹, much heralded and often cited since the digital revolution, has also not become a reality yet. In many places, work is still determined by files, folders and printers. ›Informal‹ emails are printed out on A4-paper and archived, word processing software displays the familiar proportions of a sheet of paper on the monitor, ready to be filled with pixel letters. The interplay of font and text, of haptic and materiality, as well as the recognition of universal forms in and on a ›sheet of paper‹ transform it into a ›document‹. Standardised ISO formats influence our perception of what is official and what is informal. As documents circulate, numerous other standards come into play, interlocking rather like cogwheels, making daily life easier. This usually occurs without us even being aware of it – but just imagine your ID or credit cards would no longer fit in the wallet you have just bought. Thanks to the internationally valid standard ISO/IEC 7810, though, ID documents and credit cards also always have the same proportions. The ID-1 format, for example, specifies the size of the classic banking cards, giving them an aspect ratio of 1.586:1. So although some institutes may issue you with a golden credit card, this is still not the same as owning a golden rectangle...

DIN Norm DIN 476 (equivalent to EN ISO 216) from A0 to A8

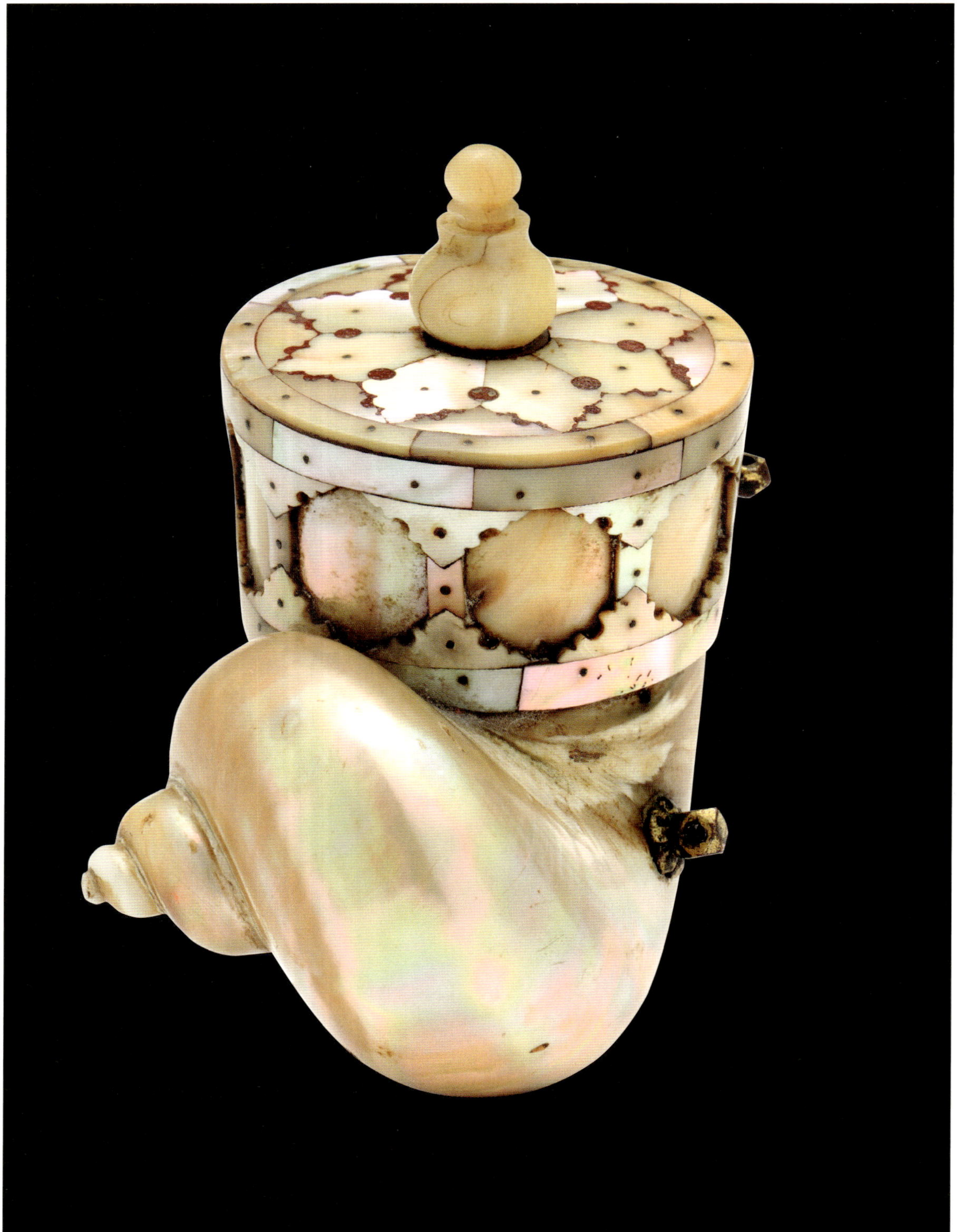

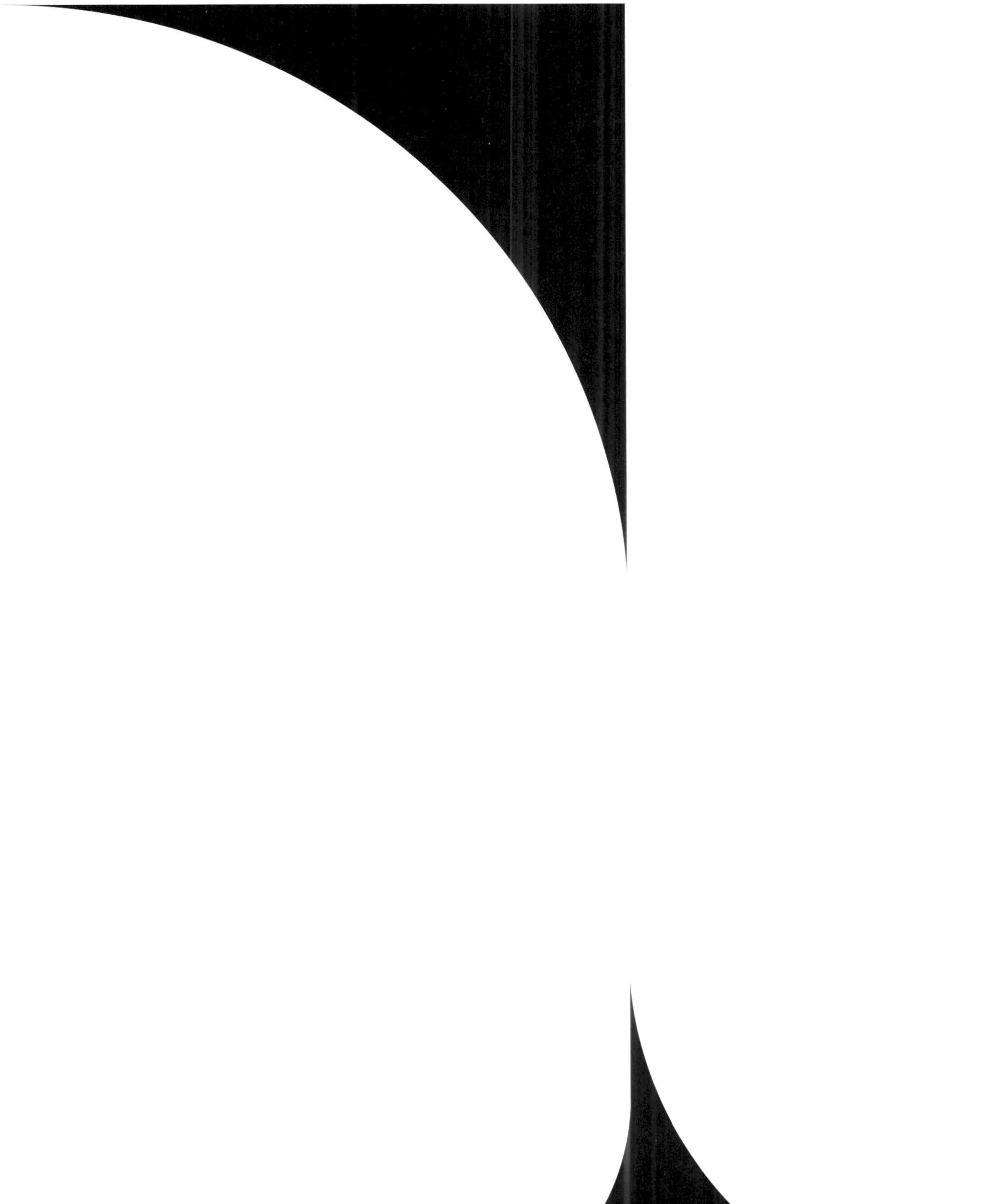

Turbo marmoratus powder horn,
Gujarat, seventeenth century

5 Theses
Why I use the golden ratio for design
Erik Spiekermann

1. The proportions are easy to remember – start with 3 and 5 and then just always add the highest previous number in the sequence, i.e. 3+5=8; 8+5=13; 13+8=21; 21+13=34, etc. But in fact as I only need a proportion and no absolute numbers, I just make a mental note of 3:5, because then I can easily divide one page by three. The ratio is also easy to remember as 10:16.

2. If necessary, it is simple to construct from the diagonal in a double square. With two horizontal double squares, draw a diagonal through the upper one and take this as the radius of a circle. The point where the circle meets the perpendicular marks the section with the larger value (e.g. 8) in relation to the height of the two double squares (e.g. 5).

Two squares from a double rectangle

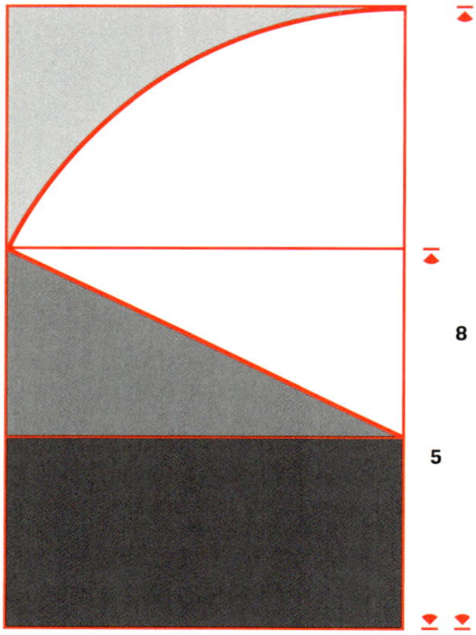

A circle with the diagonal of the rectangle as the radius

3. The German DIN standard proportions 1:1.41 (√2), i.e. 7:10, remains the same when divided in half or doubled. Practical, but boring. In contrast, the proportions of a double-page spread in the golden ratio change, so 5:8 becomes 10:8 or 5:4, virtually turning a narrow vertical rectangle into a square. When this is doubled, it again becomes the golden ratio 5:8. Moreover, it is both appealing and practical that when a square is set into a rectangle in the golden ratio, the remaining rectangle also has this beautiful proportion.

4. The narrow upright format is easy to hold in the hand, while the open double-page spread is rather wide in relation to its height (since it is 5:4), doing justice to the fact that our hands are slim while our eyes are set next to each other.

5. I find these and other instances of ›cabbalistic numerology‹ fun. They not only help me in my work, but also usually convince my clients, since no one likes to object to Golden Rules. Apart from that, I apply geometric methods to design page grids, formats and margins by drawing squares, rectangles and diagonals and so produce (in a mathematical sense following the intercept theorem) page elements with similar proportions or even the same proportions in relation to each other. To ensure this cabbalism is also practical and fits in with pre-given formats (e.g. DIN standards or column divisions and page grids with 9 or 12 divisions), one can also round up the numbers slightly and deviate from pure theory. But the golden ratio is certainly a good place to start.

Transverse rectangle 8:5 and 5:3

Rectangle 5:8 with square inside

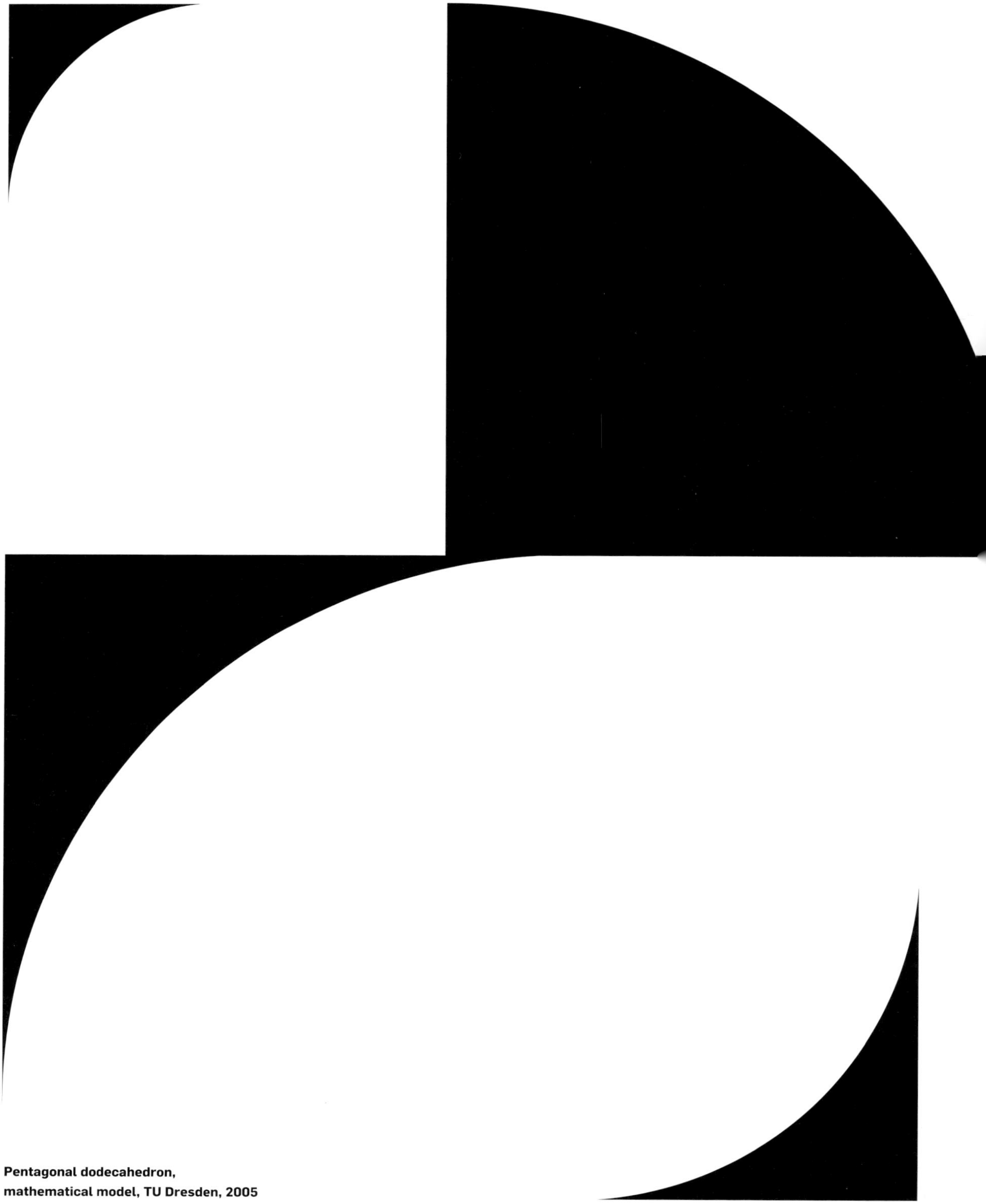

Pentagonal dodecahedron,
mathematical model, TU Dresden, 2005

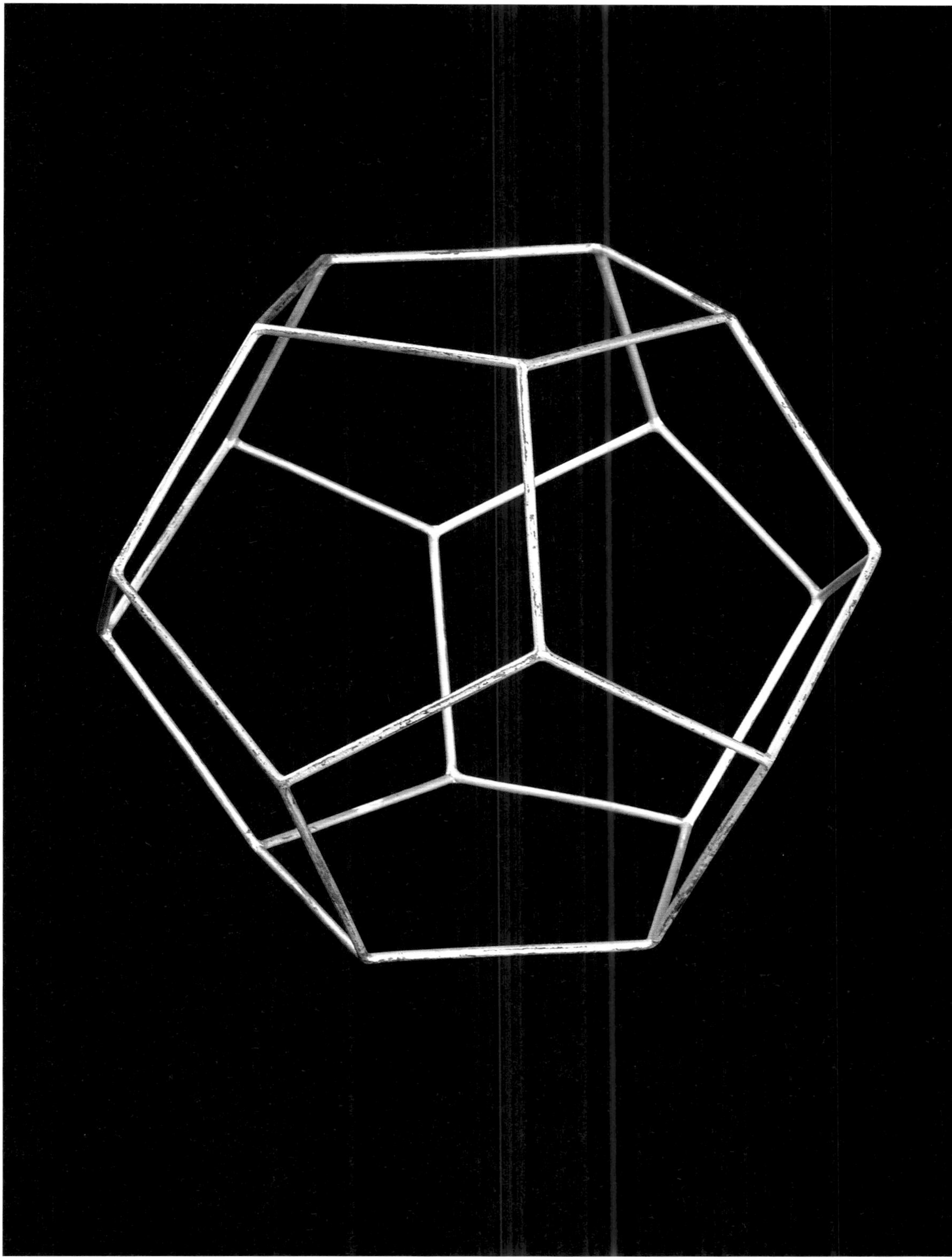

Beautiful, versatile, irrational
The mathematical beauty of the golden ratio

Albrecht Beutelspacher

The golden ratio has its roots in geometry. As a number, it takes us to completely new realms. Like practically no other value except for Pi, the golden ratio links geometry and numbers.

What is the golden ratio?

The first recorded definition of what is now called the golden ratio can be found in Euclid's famous treatise *Elements* (c. 300 BC). In Book II, Proposition 11 described the following task: »To cut a given straight line so that the rectangle contained by the whole and one of the segments equals the square on the remaining segment.« Today, it is difficult for us to understand a verbal definition of this kind, but it was a way of describing a ratio without variables and equations. If we call the length of ›one section‹ m and that of the ›other section‹ M, then today we can express the condition in the above definition this way: $(M+m)\times m = M^2$. If we divide this equation by $M\times m$, we obtain $(M+m)/M = M/m$. This leads to the definition commonly accepted today: a point S divides a segment in the golden ratio if the ratio of the whole segment to the longer part is equal to the ratio of the longer part to the shorter; in other words, if $(M+m)/M = M/m$ (Fig. 1, p. 52).

The naming of the variables here is well established. M, standing for ›major‹, denotes the length of the longer section, while m, for ›minor‹, is the smaller section. A question that inevitably arises at this point is: does a dividing point like this even exist? The answer is yes, as can be shown in at least two ways.

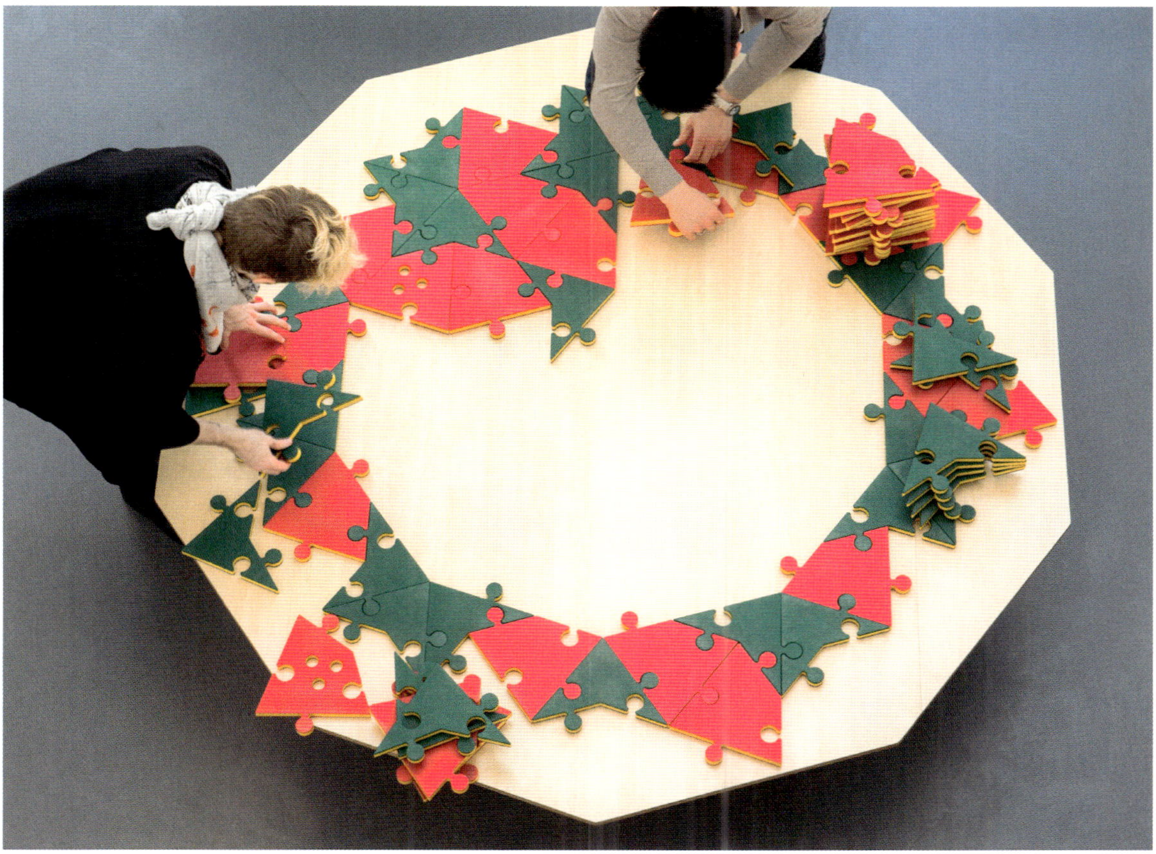

Penrose puzzle at the Mathematikum science museum in Gießen

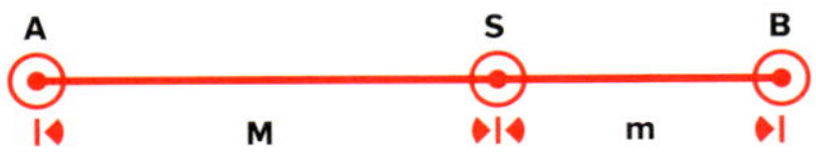

Figure 1

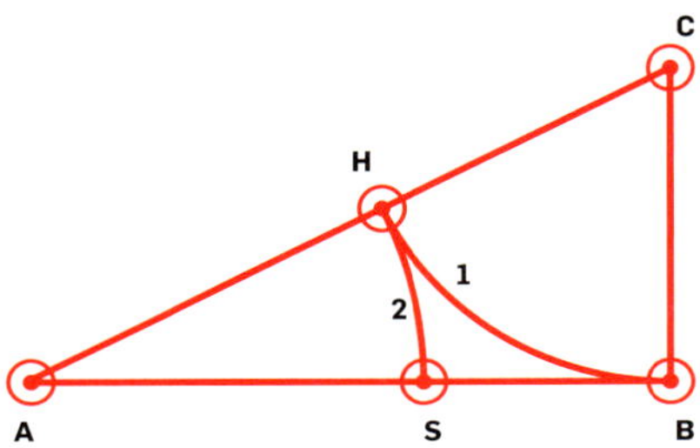

Figure 2

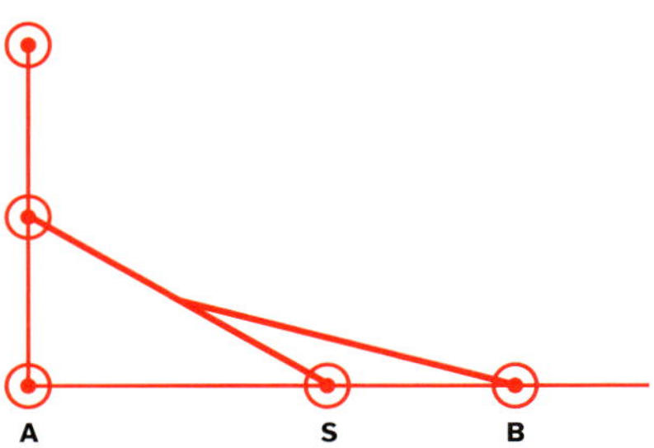

Figure 3

A. *Through a continuity argument*. We imagine that the point S moves along the segment from A to B. At every moment, we can calculate the ratios $(M+m)/M$ and M/m. When S is at the far left, M is very small and m is very large, so the left side of the equation, $(M+m)/M$, is larger than the right, M/m. On the other hand, when S is close to the end of the line to the right, M is huge and m is tiny, and $(M+m)/M$ is therefore only slightly larger than 1. However, M/m is a very large number, and thus it is certain that $(M+m)/M < M/m$. All in all, we can conclude that there must be a point ›in between‹ at which the two ratios are equal.

B. One can also precisely *construct* the point S. The standard construction is as follows (Fig. 2):

To divide the segment AB of length a in the golden ratio, construct a perpendicular segment BC of length $a/2$ at point B. The circle with centre C through B intersects the segment AC at the point H. Finally, the circle with centre A and radius AH intersects the segment AB at the point S.
This point S divides AB in the golden ratio. (Since AC has the length $\sqrt{a^2 + (a/2)^2} = (a/2)\sqrt{5}$, AH and therefore AS have the length $(a/2)\sqrt{5} - (a/2) = a/2\,(\sqrt{5} - 1)$. One can then calculate the two ratios of segment lengths and verify that they are equal.)

The artist Jo Niemeyer (born 1946), in whose works the golden ratio plays a central role, discovered the following simple construction (Fig. 3):

Three segments of equal length are required. Position the first one vertically, and mark its midpoint. Place the second segment so that its top end rests on that point. Similarly, place the third segment so that its top end rests on the midpoint of the second. Then the points at which the bottom ends of the segments meet the underlying horizontal line have the property that the middle point divides the other two in the golden ratio.
(Using the Pythagorean theorem, one can calculate the segment lengths along the horizontal line and verify that the two ratios arising from the division point are equal.)

The numerical value of the golden ratio, for example, the number M/m, is often designated with the Greek letter ϕ (phi).

Where does the golden ratio appear?

The first known definition of the golden ratio can be found in the writings of Euclid, but the ratio had actually been discovered centuries earlier. For the Pythagoreans, as the school of Pythagoras (c. 500 BC) is known, it was both a symbol and a scandal. They discovered the golden ratio in the regular pentagon and the pentagram. But before we turn our attention to the Pythagoreans, let us first present some basic facts about the regular pentagon.

We call a convex pentagon *regular* if its sides are equal in length and its angles are equal. Now consider a regular pentagon with a side length of a.

A segment that connects two non-adjacent corners of a regular pentagon is called a *diagonal* (Fig. 4).

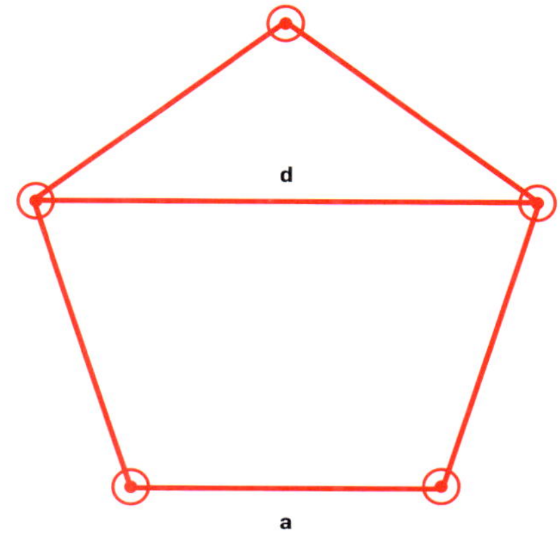

Figure 4

Every diagonal can be combined with two adjacent sides of the pentagon to form an isosceles triangle. All these triangles are congruent according to the SAS (side-angle-side) congruence theorem. In particular, therefore, all diagonals have the same length d; let α be the base angle of these triangles.
Furthermore, all the isosceles triangles that consist of two diagonals intersecting at a corner of the pentagon are also congruent (side-side-side – SSS, Fig. 5).

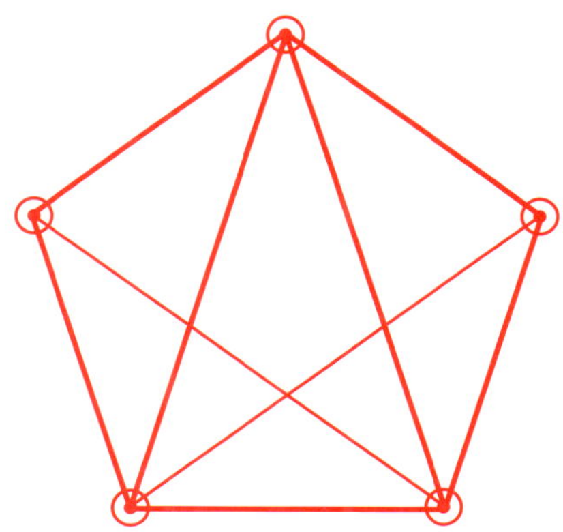

Figure 5

If we designate the angle at the apex as β, then the size of every base angle in one of these triangles is $\alpha+\beta$. Thus, according to the theorem regarding the sum of the interior angles of a triangle, $2(\alpha+\beta) + \beta = 180°$.

Now a triangle consisting of a diagonal and two adjacent sides has an angle measuring $\alpha+\beta+\alpha$ at its apex. Hence, the theorem on the sum of the angles of a triangle states that $\alpha+\beta+\alpha + 2\alpha = 180°$. When the two equations are combined, the result is $5\beta = 180°$, or $\beta = 36°$. It then also follows that $\alpha = 36°$.

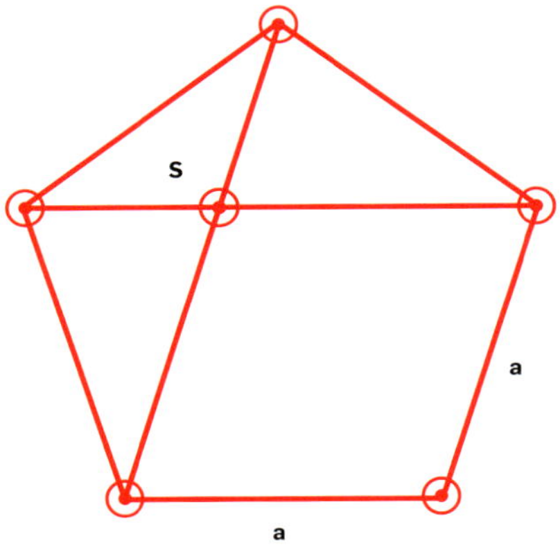

Figure 6

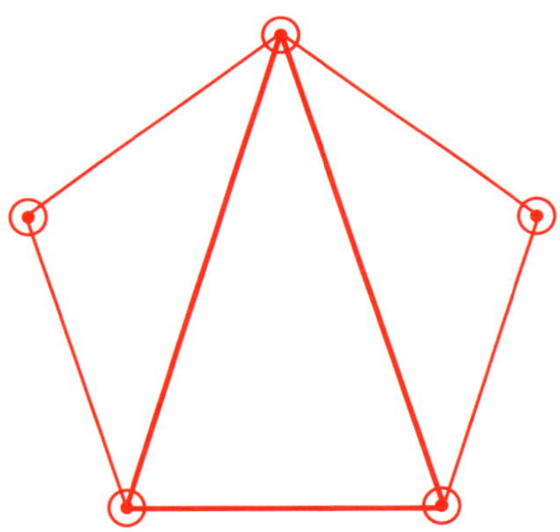

Figure 7

Now we can conclude that each diagonal is parallel to the edge that lies opposite from it, since the sum of the angles at the edge meeting each of those segments is $2\alpha+\beta + \alpha+\beta = 5\alpha = 180°$.

Thus, every pair of diagonals intersecting at a point inside the pentagon form a rhombus (Fig. 6). (From the previous conclusion, it follows that they form a parallelogram, but since two adjacent sides have the same length a, this is a rhombus.)

Let us now consider two diagonals that intersect at an interior point S. We can apply the second intercept theorem from point S and obtain $d/a = a/(d-a)$, where d is the length of a diagonal.

Interpreting this equation for an individual diagonal, d/a would be the ratio of the entire diagonal to the larger part, and $a/(d-a)$ the ratio of the larger part to the smaller. In other words, the point S divides the diagonals in the golden ratio.

We can now summarise these results in the following theorem: *in a regular pentagon, any two diagonals that have no corner in common intersect at a point dividing them in the golden ratio.* In other words, the golden ratio is the ratio of diagonal length d to side length a.

This theorem can also be formulated as follows: in a pentagram, every ›inside corner‹ divides the segment that contains it, from outer vertex to outer vertex, in the golden ratio.
As it happens, one can also use this insight to construct a regular pentagon. Given a segment AB of length a, construct a circle of radius ϕa around each end point. One of the two points of intersection of the circles is another corner of the pentagon with side AB. It is then easy to construct the two remaining corners (Fig. 7).

The first irrational number

The remarkable thing is that the mathematicians of the Pythagorean school also discovered that the golden ratio is an irrational number. Not only was this the first irrational number discovered in the history of mathematics, but it was, in fact, the golden ratio that made us aware that irrational numbers exist at all! Hippasus of Metapontum (c. 500 BC) is often credited with having discovered ›incommensurability‹, although recorded evidence to support this claim is lacking.

More precisely, then, the Pythagoreans discovered that the ratio of the diagonal length and the side length of a regular pentagon cannot be expressed as a fraction of integers; it is not a rational number. In other words, this ratio is not rational but ›irrational‹.

The proof of this is ingenious and testifies to a supreme mastery of mathematical techniques. Its external structure is that of a proof by contradiction. We assume that the ratio d/a of diagonal length d and side length a of a regular pentagon is a rational number and can therefore be represented as a fraction. The ratio might then be a fraction like 8/5, for example. In such a case, d would consist of 8 and a of 5 units of length. To put it another way, both d and a are natural numbers. Now, out of all the regular pentagons with integral diagonals and side lengths, we select the smallest one, the one with the smallest d, for example. With all the groundwork has been laid, we can now turn to the actual proof, and show that a regular pentagon exists with smaller integral sides and diagonal lengths.

To that end, we note that the diagonals of a regular pentagon enclose a smaller pentagon, and show that this pentagon likewise has integral side and diagonal lengths.

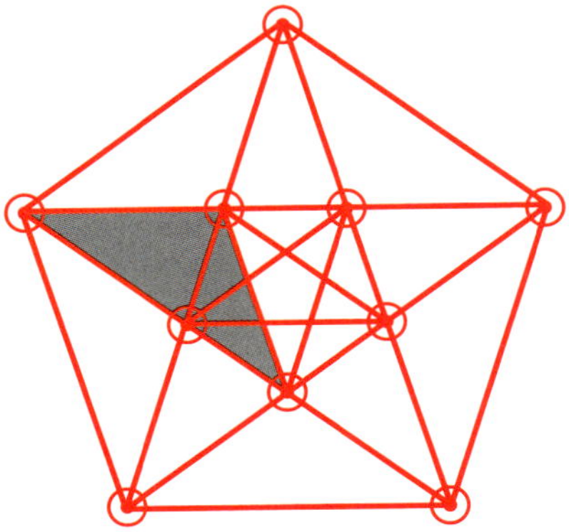

Figure 8

n	1	2	3	4	5
f_{n+1}/f_n	2	1.5	1.667	1.6	1.625

Figure 9

The side length of the small pentagon is $2a - d$, in other words an integer. To obtain the diagonal length, we consider the shaded triangle. Since its base angles are equal, this is an isosceles triangle. The length of a diagonal of the small pentagon is therefore $d - a$, which is also a natural number and, moreover, one that is smaller than d (Fig. 8).

We have thus found a regular pentagon with integral diagonal and side lengths that is smaller than the initial pentagon – and contradicting the assumption that d was the minimum diagonal length.

The number ϕ

If the defining equation $(M+m)m = M^2$ for the golden ratio is divided by m^2 and solved for the unknown ratio M/m, one obtains $(M+m)/m = M^2/m^2$, and therefore $M/m + 1 = (M/m)^2$ or $(M/m)^2 - M/m - 1 = 0$. Hence, the value of the golden ratio is a solution of the equation $x^2 - x - 1 = 0$. The solutions of this quadratic equation are $1 + \sqrt{5}/2$ and $1 - \sqrt{5}/2$. As the ratio of the lengths of two line segments, the golden ratio is a positive real number, and so only the first solution is relevant, which is $M/m = 1 + \sqrt{5}/2$. The symbol ϕ is also used to designate this number.

When the approximate value is calculated, one obtains $\phi \approx 1.618$.

If we rearrange the equation $x^2 - x - 1 = 0$, we obtain $x^2 = x + 1$ and $x(x - 1) = 1$. This can be interpreted as follows: $\phi^2 = \phi + 1$ (›one obtains ϕ^2 by adding 1 to ϕ‹) and $\phi(\phi - 1) = 1$. The last equation says that $\phi - 1$ is the number which, when multiplied by ϕ, yields 1. In other words, $\phi - 1$ is the multiplicative inverse of ϕ, which can be expressed as $\phi - 1 = \phi$-1 (≈ 0.618).

This calculation provides a practical rule for identifying the golden ratio. A segment is in the golden ratio to a subsegment if the latter covers about 61.8 %, or just over 60 %, of the whole.

The Fibonacci connection — or the golden ratio made easy

In 1202, Italian mathematician Fibonacci, also known as Leonardo of Pisa (c. 1170–1241), wrote his *liber abaci* (Book of Calculation) which, among other things, also introduced what later became known as Fibonacci numbers. At that time, though, he had no idea just how important this number sequence would become in general, or its close connection to the golden ratio.

The Fibonacci numbers form the integer sequence 1, 2, 3, 5, 8, 13, 21, 34, 55 ..., beginning with 1 and 2, where any number is the sum of the two preceding numbers. The Fibonacci numbers would thus continue as follows: 89 (= 34+55), 144 (= 55+89), etc. We designate the nth Fibonacci number as f_n; for example, $f_5 = 8$.

The Fibonacci numbers have many wonderful properties, but the key one in relation to the golden ratio is that the sequence of quotients f_n+1/f_n of the Fibonacci numbers converges to the golden ratio. The table (Fig. 9) shows just how quickly this happens:

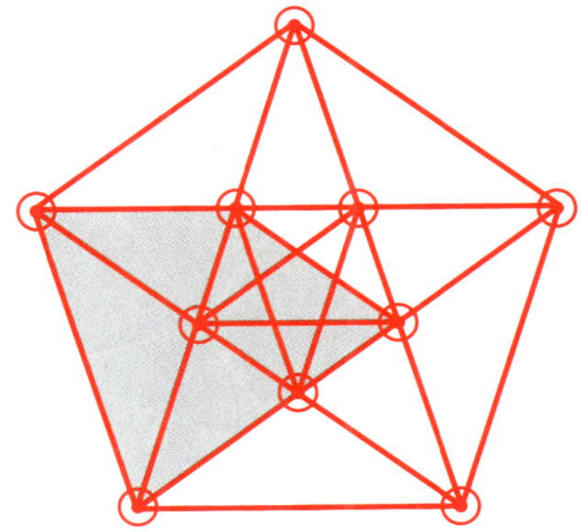

Figure 10

Penrose tiles — or how to master infinity with the golden ratio

Tiling refers to covering a plane with ›tiles‹ without any gaps or overlapping regions. The best-known tilings are *periodic*, which means that one can obtain the whole tiling by shifting one tile or a finite number of tiles in two directions (infinitely often). This is a wonderful method of covering the infinite plane, since you only have to know a finite part in order to deal with everything.
One can also ask whether other tilings exist, i.e., ones for which there is no translation (apart from identity) that causes the tiling to repeat. Tilings of this kind are called aperiodic. To put the question more precisely, are there sets of tiles that allow only *aperiodic* tilings?

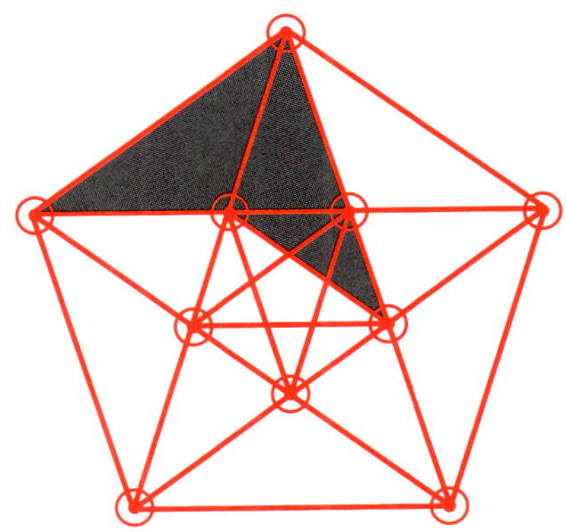

Figure 11

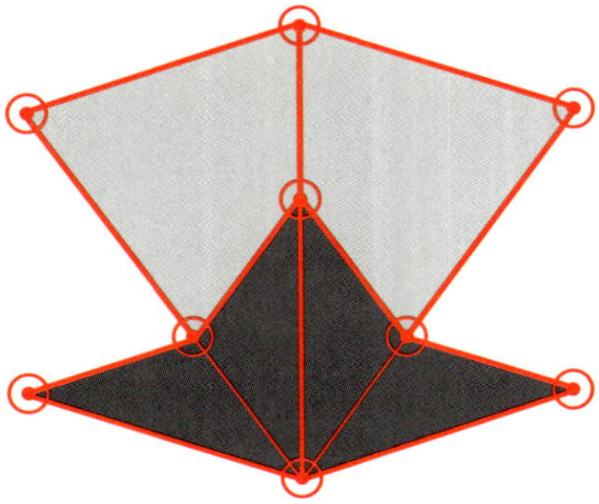

Figure 12

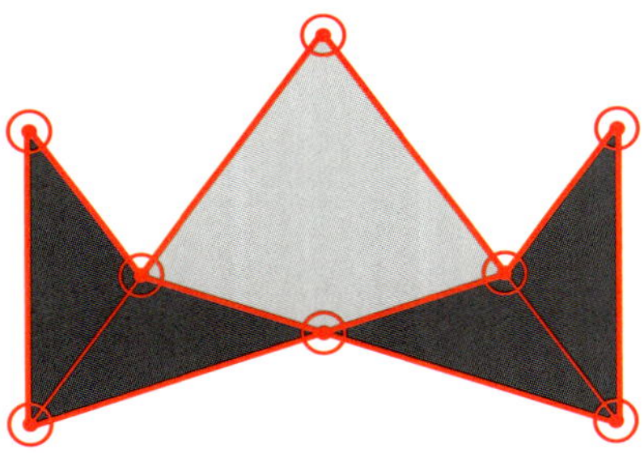

Figure 13

The answer is yes, and the most famous such tiling was produced by Roger Penrose (1974). This tiling is closely linked to the golden ratio in a variety of ways. On the one hand, the two tiles, namely the *kite* and the *dart*, arise from a regular pentagon (Fig. 10/11, p. 57).

The kite and dart each have sides of lengths a and ϕa. If you were to piece them together as the construction suggests, you would obtain a rhombus able to tile the plane periodically. Penrose then invented rules for laying the tiles to prevent the formation of shapes able to tile the plane periodically. These rules can be realised in physical tiles by creating appropriate indentations and protrusions on the kites and darts.

On the other hand, there is another, much deeper link between the golden ratio and the Penrose tiles, since if you take a naive approach to laying these tiles, you will quickly reach a dead-end and fail to produce a tiling. So the question is whether one can obtain a tiling at all – and again, the answer to this question is yes.

The method is as follows: one begins with a ›partial tiling‹ (a partial covering of the plane without any gaps), such as a single tile. Then one enlarges this partial tiling by extending each length by the factor ϕ. The area is thus enlarged by the factor ϕ^2 - a process called *composition*. Then, one covers the larger area with kites and darts of the original size. The following pictures illustrate how this works (Fig. 12/13).

A large kite decomposes to two small kites and two small half-darts. A large dart, on the other hand, can be decomposed to a small kite and two small half-darts (Fig. 12).

The rules for laying the tiles are defined in such a way that each second half of the small darts always combine with the second half of adjacent small darts to form a small dart –a process called *decomposition* (Fig. 13).

By repeating the processes of composition and decomposition, arbitrarily large areas can be covered. Ultimately, every point in the plane is covered. Hence, one does in fact obtain a tiling in this way. But why are these tilings aperiodic?

This can be demonstrated with a very sophisticated argument in which the golden ratio plays the covert leading role. If the tiling were periodic, there would be a basis set of n kites and m darts that would allow one to construct the entire tiling with the help of translations in two directions. Then, the ratio of kites to darts in the whole tiling would also be n/m, a rational number (a fraction of integers).

On the other hand, if one starts with a kite and performs the composition and decomposition steps, then after the nth step, one has f_{2n} kites and f_{2n-1} darts (where $f_1 = 1$, $f_2 = 2$, $f_3 = 3$, $f_4 = 5$, ... are the Fibonacci numbers). The table illustrates this (Fig. 14).

To calculate these numbers, consider that in each step, one kite produces two kites and a dart (more precisely, two half-darts). Similarly, in each step, one dart gives rise to a kite and a dart. Consequently, in the nth step the ratio of kites to darts is f_{2n} / f_{2n-1}. At the limit, when the tiling of the plane is complete, the ratio is equal to the limit of the quotient of successive Fibonacci numbers, which is ϕ. Thus, in a tiling created in this way, the ratio of kites to darts is equal to the golden ratio.

However, since ϕ is an irrational number, the tiling cannot be periodic.

n	1	2	3	4	5	6	7
kites	1	2	5	13	34	89	233
darts	0	1	3	8	21	55	144

Figure 14

Beutelspacher, Albrecht/Petri, Bernhard: Der goldene Schnitt, Heidelberg/Berlin/Oxford, Spektrum Akademischer Verlag 1996.
Beutelspacher, Albrecht: Wie man in eine Seifenbase schlüpft. Die Welt der Mathematik in 100 Experimenten, Munich, C. H. Beck 2015.
Walser, Hans: Der Goldene Schnitt, Leipzig, Edition am Gutenbergplatz 2013.

In equal parts
Symmetry

Florian Schütz

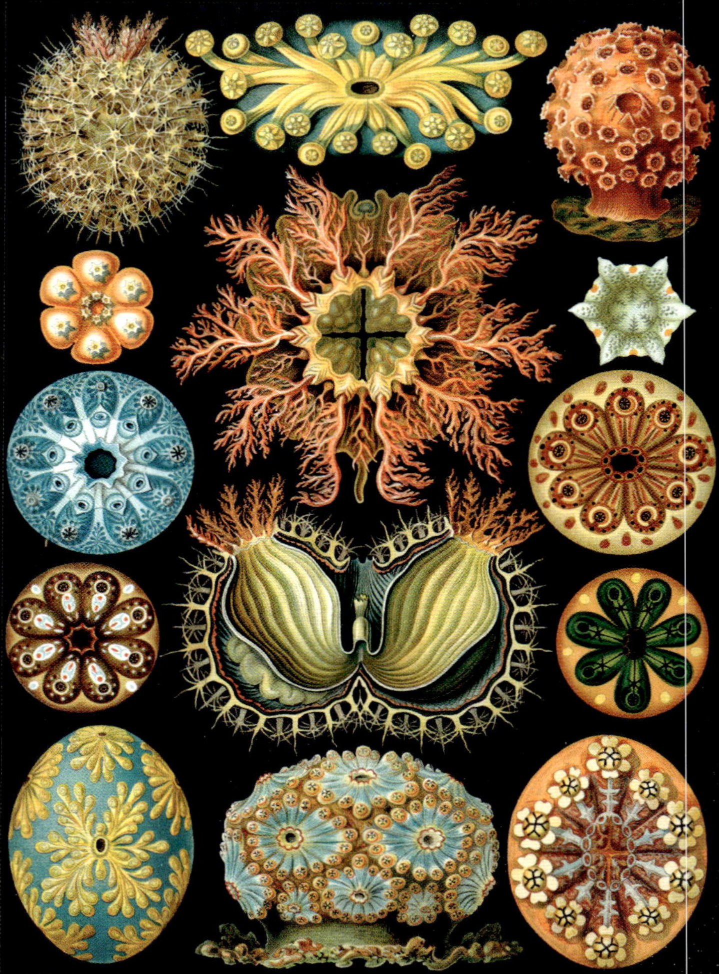

Symmetrical *Ascidiacea* (sea squirts) from Ernst Haeckel's *Art Forms in Nature* (Kunstformen der Natur), 1904

Symmetry, which is derived from the classical Greek word *symmetria*, describes the ›harmony‹ of a geometric object. A given geometric figure is regarded as having ›due and just proportions‹ or being ›well-balanced‹ if, through movement, the parts are congruent – for example, form a mirror image when folded on a line of symmetry. Among other things, symmetry is considered as a ›beautiful‹ proportional relationship: a symmetrical face, a symmetrical building or a symmetrically structured painting. However, as always in such cases, things are not quite that simple.

First of all, there are different kinds of symmetry. The most intuitive form is probably line symmetry, mentioned above. In line symmetry, also known as axial symmetry, the two halves of a figure divided along a single axis correspond exactly. The axis of symmetry, for instance, can be created by drawing a vertical line through the mid-point of the opposite sides of a square. In contrast, point symmetry describes the symmetry of a figure around a point rather than a line – for example, two circles of the same radius set next to each other and touching at one point. Here, a distinction is also made between two-dimensional and three-dimensional figures.

Not only is the origin of the word ›symmetry‹ rooted in ancient Greek, but classical Greek architecture is hardly conceivable without symmetry. In the case of the Acropolis in Athens, the line symmetry of the front façade's columns and decorative pediment is only too obvious. Moreover, classical Greek architecture rarely employed an uneven number of columns. Prestigious display or sacred buildings in other cultural regions have similarly been inspired by the idea of symmetrical design, from the Taj Mahal in Agra to the Brandenburg Gate in Berlin and the many villas designed by Italian Renaissance architect Andrea Palladio. Such a list could be continued almost indefinitely. Even though the internal space in structures employing a symmetrical front façade can be divided in many different ways, the many examples of such symmetrical designs may well be the most powerful argument for a conscious application of a particular proportional ratio.

Not only do such monumental, seemingly perfect structures, carefully designed down to the smallest detail of their ›over-powering‹ architecture, function as a reminder to onlookers and outsiders of their own transitory lives, but they are also intended to show off the status and skills of their builders and residents. Yet as Madame de Maintenon noted, less than enamoured with life in the magnificence of the Palace of Versailles with its almost over-fastidious landscaped grounds, ultimately one could »die from symmetry«.

As is evident from Madame de Maintenon's remark, symmetry does not always covey harmony, sublimity or even divinity. While symmetrical architecture can certainly be impressive, ›well-proportioned‹ symmetrical images can also confuse or puzzle beholders, suggest (unconscious) tensions and disturb the dynamics of a depicted scene. Albrecht Dürer's *Self-Portrait* from 1500, executed in a symmetrical frontal view, may include clichéd features recalling an idealised depiction of Christ, yet all ›great‹ paintings, live from their dynamism, changing perspective and challenge to beholders to embark on a voyage of discovery. Dürer's painting certainly does more than justice to such a challenge, and viewers can find the golden ratio in this symmetrically structured composition – although there is no evidence that Dürer intentionally used this device or that it was part of his artistic repertoire.

In general, symmetry – in common with the golden ratio – does not automatically produce harmony. In biology as well, where the human body at first glance seems to have an axial symmetry from the top of our heads to the soles of our feet, the details are subject to different systems. A totally symmetrical face awakens a feeling of unease – as is evident if we use a mirror to double just one side of our own face. Instead, in real life, one ear may be larger and, as anatomy tells us, some organs are only located on one side of our bodies.

In such areas as architecture, biology or art, symmetry often takes its place as one proportional ratio among many. But it is not an all-embracing principle of order with inherent magical qualities. Symmetry is not the holy grail of harmony. It can be a stylistic device, a distinguishing feature. Through his great love of symmetrical compositions, film director Wes Anderson has created his own distinctive visual style, now iconic for his movies – probably also because their unusual and slightly bizarre style make them so memorable for viewers.

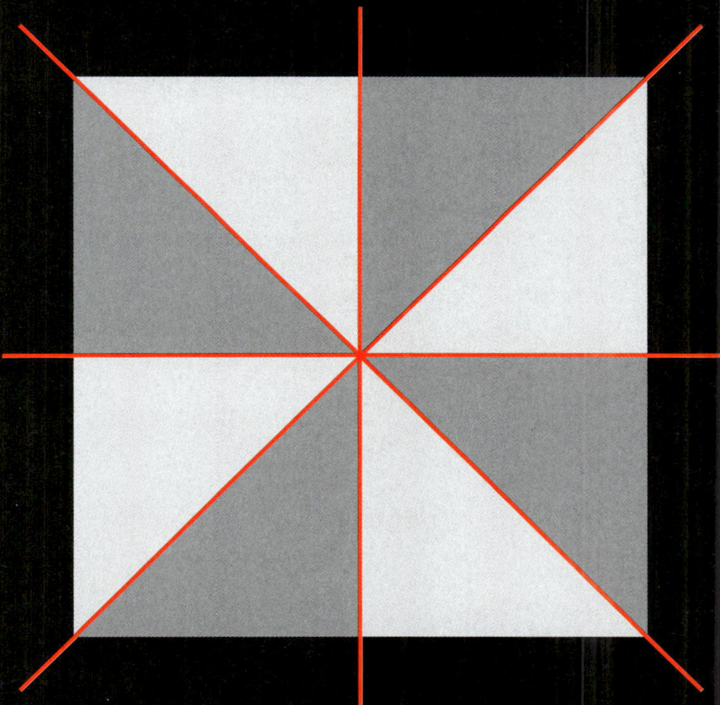

Lines of

Le Corbusier's system of proportions
The Modulor

Bärbel Högner

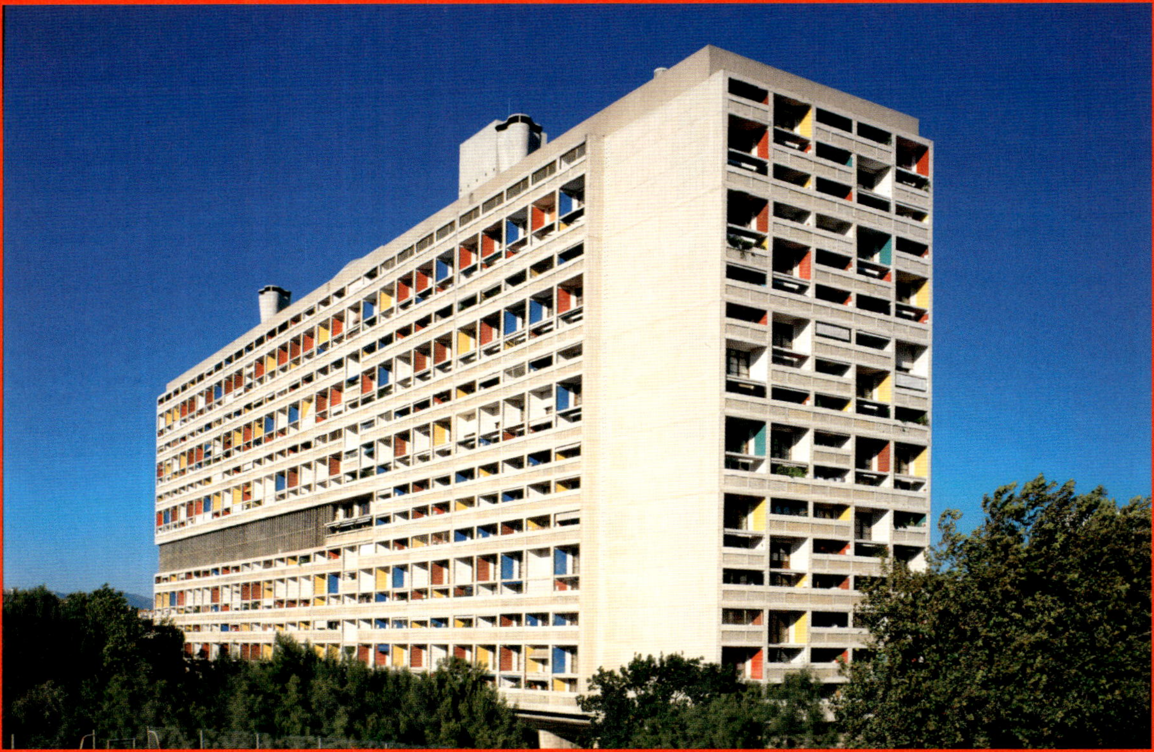

Figure 1. ***Unité d'Habitation*** **in Marseilles**

Le Corbusier's *Unité d'Habitation* in Marseille (Fig. 1) is an icon of modernist architecture. Its wedge-shaped pilotis (concrete legs) support over three hundred maisonette apartments stacked above and adjacent to each other like bottles in a rack. The seventeen-storey residential housing project, completed in 1952, has a corridor of shops as well as its own kindergarten. This ›vertical city‹ is crowned by a roof landscape including a sports hall, swimming pool and an open-air theatre. Commissioned by the French Ministry of Reconstruction and Town Planning in 1946, this was Le Corbusier's largest project at that time. Here, he first tested various design elements that were characteristic of his later work. The external appearance of the residential block, regarded as an example of brutalist architecture, is impressively influenced by Le Corbusier's use of *béton brut* (fair-faced concrete) and the systematic application of his *polychromy* (colour keyboards). At first glance, it is less obvious that he also applied an innovative system of proportions. In fact, though, his Modulor scale determines the dimensions of the building and the layout of the apartments. In 1947, Le Corbusier publicly presented his Modulor invention – two sequences of numbers known as the ›red and blue series‹ – at New York's MoMA. His unconventional values were based on set points on the human body, with the rhythm of intervals following the golden ratio and the principle governing the Fibonacci sequence.[1]

Diversity instead of standardisation

Working together with his team in his atelier, Le Corbusier developed the Modulor, his ›miracle of numbers‹ between 1943 and 1946.[2] The graduated linear series, with diminishing intervals towards zero and enlarging intervals as the sequence progresses towards infinity, was intended to provide potential combinations always reliably guaranteeing the harmonious design of lengths, surfaces and volumes.[3]

In his first publication on the Modulor (from 1950), Le Corbusier included many autobiographical notes to show how he had been constantly preoccupied with the subject of modulatory lines and the question of harmonious surface ratios throughout his life.[4] In developing a varied and universally applicable system of proportions, Le Corbusier had two concerns in mind. First, he hoped to replace the abstract unit of measurement in the metric system with a graduated scale of intervals which, since it was derived from key values based on the human frame, would be a ›human scale‹.[5] Second, he envisaged this as a unified system employed the world over since, in his view, the two common standardised units of the ›inch and foot‹ and the ›metre‹ were relatively incompatible.

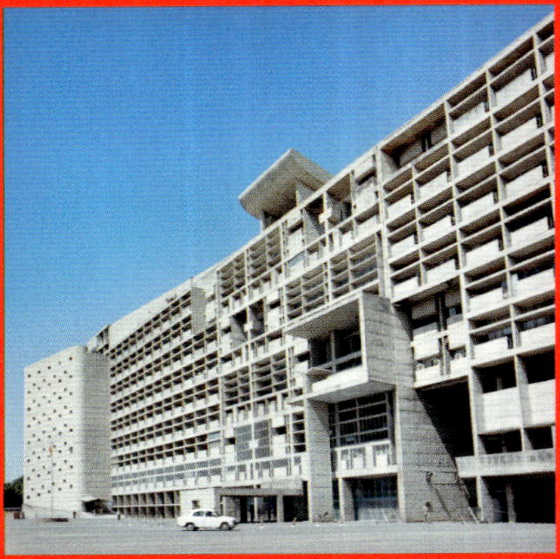

Figure 2. Administration building in Chandigarh, India

Early calculations, sketches and experiments assumed a standard body height of 175 centimetres. In 1945, the first Modulor measuring tape was produced and, given the results, Le Corbusier talked of this as a ›proportional regulator‹. On his way from Le Havre to New York on a cargo ship, the voyage, described as hazardous, took nearly three weeks due to the rough and stormy seas. However, this did then offer a kind of retreat where Le Corbusier again had to time to dedicate himself fully to coordinating the graduated scales, ensuring they were defined as truly »anthropocentric«.[6] Once back in France, he then tried to find a succinct and catchy name for this system. Ever since the Modulor brand name was created by combining *module* and *or* (as in ›gold‹), it has been represented by the iconic stylised contour of a broad-shouldered figure with one arm raised, and with a markedly large hand. The issue of the compatibility between metric and the customary imperial and US systems of measurement still remained, but this was resolved by recourse to the supposed height of the leading male protagonists in British detective stories since, as one of Le Corbusier's assistants pointed out, the detectives were often six foot tall. With the figures adjusted minimally upwards and rounded off, the graduated scale with body height at the fixed point of 183 centimetres could be transferred to international customary systems.[7]

In summing up the development in 1950, Le Corbusier noted that his office had researched and experimented for a period of seven years until the two scales were practicable for everyday use – and, he added, the resulting Modulor could be learned by anyone in just five minutes.[8] The exercise used the following data, beginning with the basic unit of a square with sides measuring 113 centimetres. In relation to the specified body height of 183 centimetres, this equals the height of the navel. Adding a second square to the first produces a total side length of 226 centimetres – precisely the spatial displacement created by Le Corbusier's ideal person of 183 centimetres standing with one arm raised. A third square is then inserted at the set points of the golden ratio on both side lengths – at 43:70 as well as 86:143 centimetres. This third square then »determines the location of the right angle«. This passes through the diagonals of the two basic units, creating two points of intersection in the third square. A straight line set over these points of intersection produces the ›diminishing‹ and ›increasing‹ series – the Modulor's red and blue series of numbers respectively.[9] (see Fig. 4)

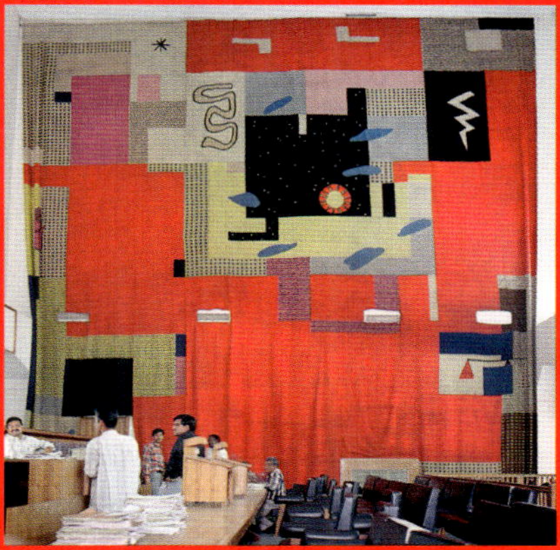

Figure 3. Wall hanging with a pattern based on the Modulor, Chandigarh, India

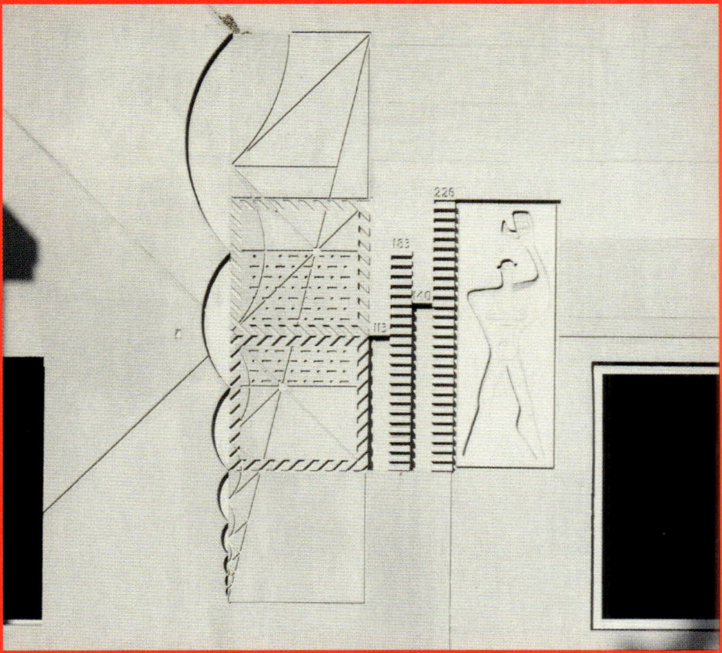

Figure 4. Façade of the *Unité d'Habitation* in Berlin

Logical implementation, varied reception

After completing the block of flats in Marseilles, Le Corbusier also designed all his subsequent projects with the help of the Modulor. He applied it most extensively from 1951 in his work on the master-plan for the city of Chandigarh in northern India, a major project involving Le Corbusier as an urban and landscape planner right up until his death. Yet whether he was designing the parliament building, the High Court of Justice, the art college or the museum, the ground plans and façades (Fig. 2, p. 66) always followed his new scale of dimensions. Its visual qualities are also evident from the design of tapestry wall hangings (Fig. 3, p. 67). The Modulor was even taken as the basis for the proportions of the ›Open Hand‹ monument created for the city.[10]

Le Corbusier's first attempts at creating a rhythmic composition of ribbon windows in Chandigarh were then transferred to France. For this, he turned to Iannis Xenakis, at that time an engineer in his architect's atelier, and asked him to refine the glazing system from India. For the La Tourette monastery, Xenakis, who was also a composer, created the famous *pans de verre ondulatoires* (undulating glass surfaces) for the structure of strip floor-to-ceiling windows set in the façade following divisions based on Modulor proportions. The *crescendo* and *decrescendo* of their grid pattern resonated with one of Le Corbusier's early dreams – the longing to find a body of rules for proportions in the style of a musical scale.

The reaction to the Modulor was reserved, despite Le Corbusier's substantial efforts to promote his universally applicable invention and his vision of an association of friends of the Modulor active around the globe.[11] In the early 1950s, the Swiss journal *Werk* was still optimistic that the Modulor would expand the potential of architectural expression.[12] Not long afterwards, Le Corbusier published a second book on his system of proportions (first edition 1955), offering a wealth of evaluations from a circle of experts as well as personal examples of its use in practice. Some architects did attempt to apply the Modulor, but it seems that many would agree with Richard Meier when he said that he

1 Based on the Modulor scale each apartment in Marseilles measures 3.66 metres wide and 2.26 metres in height. Compare Jean-Louis Cohen 2004, pp. 56–59 and Frampton 2001, pp. 154–165.
2 Le Corbusier 2003a, p. 92.
3 The trigger came from the efforts of the French state to introduce a standardisation system for the period of reconstruction after the Second World War. Le Corbusier was concerned, he feared that it would produce a »poor« method (Le Corbusier 2003a, p. 36). Moreover, his fellow French architects were involved in an intensive debate over proportions. On this, see Cohen 2014, pp. 1–14.
4 Le Corbusier noted that as a student, he had already realised that there was a correlation between the diversity of forms and order in nature. Later his paintings had served as a vehicle to express his fascination with geometry, and he had used the right angle and golden ratio as a means to structure surfaces (Le Corbusier 2003a, pp. 25–27).
5 The models were old measures such as the cubit (elbow), finger (digit), foot and pace – mathematics which formed an integral part of the human body (Le Corbusier 2003a, p. 19).
6 Le Corbusier 2003a, p. 50. The decisive points of the human frame were defined as seat and table height, the height of the solar plexus, shoulders and the top of the head.
7 Le Corbusier 2003a, pp. 56–58.
8 Le Corbusier 2003a, p. 85.
9 Le Corbusier 2003b, here p. 43.
10 On the use of the Modulor in Chandigarh, see Le Corbusier 2003b, pp. 227–235; 269–274; 280–291.
11 Le Corbusier 2003a, p. 239.
12 Hoesli 1954, pp. 15–19.
13 On this, see Meier.
14 On the Unite d'Habitation in Berlin, see Högner 2008 and Müller-Reppen 2008.

had tried to use this dimensioning system, but had given up.[13] Nonetheless, the Modulor achieved world fame – no doubt due to Le Corbusier's persistence and willpower in developing it and his ensuing identification with it. In short, one could say: the Modulor is Le Corbusier. Today, some businesses from the world of architecture and design even bear this distinctive name – a phenomenon that can be taken as indicative of the respect for Le Corbusier's dimensioning system. In addition, the Modulor, as part of the architectural work of Le Corbusier, was inscribed as UNESCO World Heritage on July 17, 2016. With the approval of this transnational application, handed in by seven nations, 17 sites by Le Corbusier have been awarded this prestigious status. The sites include the buildings mentioned above from his late works, as well as the National Museum of Western Art, Tokyo, and the cultural centre in Firminy, France – all based on the Modulor's proportions.

In 1956, Le Corbusier was invited to design a *Unité d'Habitation* similar to the prototype in Marseilles for Berlin's Interbau 57 International Building Exhibition. Much to his annoyance, he was not allowed to use the Modulor scale for this residential block near Berlin's Olympic Stadium, since the local housing authority provisions specified a ceiling height of 250 centimetres. Residents in these public housing apartments have always appreciated the higher ceilings – too high, according to the French model – as well as the larger floor space, since the length and breadth of the apartments were also proportionally aligned.[14] But the Modulor can still be found at other places in the residential block. On the front façade of the building, flanking reliefs of the Modulor figure and drawings of the Modulor scale are set in the concrete around the entrance area (Fig. 4, p. 67). Once inside the foyer, Le Corbusier's ›magic formula‹ is omnipresent. In 1998, the floor in the apartment block, now classified as a heritage building, was refurbished in accordance with the ›master's‹ original plans. Since then, residents and guests cross the foyer walk on a pattern of slate slabs which – similarly to the tapestry wall hangings in India or the windows in La Tourette – represent the Modulor's ›game of volumes‹.

Cohen, Jean-Louis: Le Corbusier, Cologne, Taschen Verlag 2004.

Cohen, Jean-Louis: »Le Corbusier's Modulor and the Debate on Proportion in France«, in: Architectural Histories, 2(1): 23, 2014, pp. 1–14 (DOI: http://dx.doi.org/10.5334/ah.by, Accessed: 9 April 2016).

Frampton, Kenneth: Le Corbusier, New York, Thames & Hudson 2001.

Högner, Bärbel: Typ Berlin. Das Corbusierhaus in Charlottenburg, Berlin, Jovis 2008.

Hoesli, Bernhard: »Le Corbusiers Modulor«, in Werk, Volume 41, Issue 1, 1954, pp. 15–19 (DOI: http://dx.doi.org/10.5169/seals-31690, Accessed: 9 April 2016).

Le Corbusier: Der Modulor. Darstellung eines in Architektur und Technik allgemein anwendbaren harmonischen Maßes im menschlichen Maßstab, Stuttgart, DVA 2003a (Facsimile reproduction of the German translation, second edition, 1956).

Le Corbusier: Modulor 2, 1955. Das Wort haben die Benützer. Fortsetzung von »Der Modulor« (1948), Stuttgart, DVA 2003b (Reprint of the 1958 German edition).

Meier, Richard: Online interview: http://www.webofstories.com/play/richard.meier/12;jsessionid=A308E0C4C (Accessed: 12 May 2016).

Müller-Reppen, Frithjof (ed.): Le Corbusiers Wohneinheit am Heilsberger Dreieck »Typ Berlin«, Berlin, Jovis 2008 (Reprint of the 1958 edition).

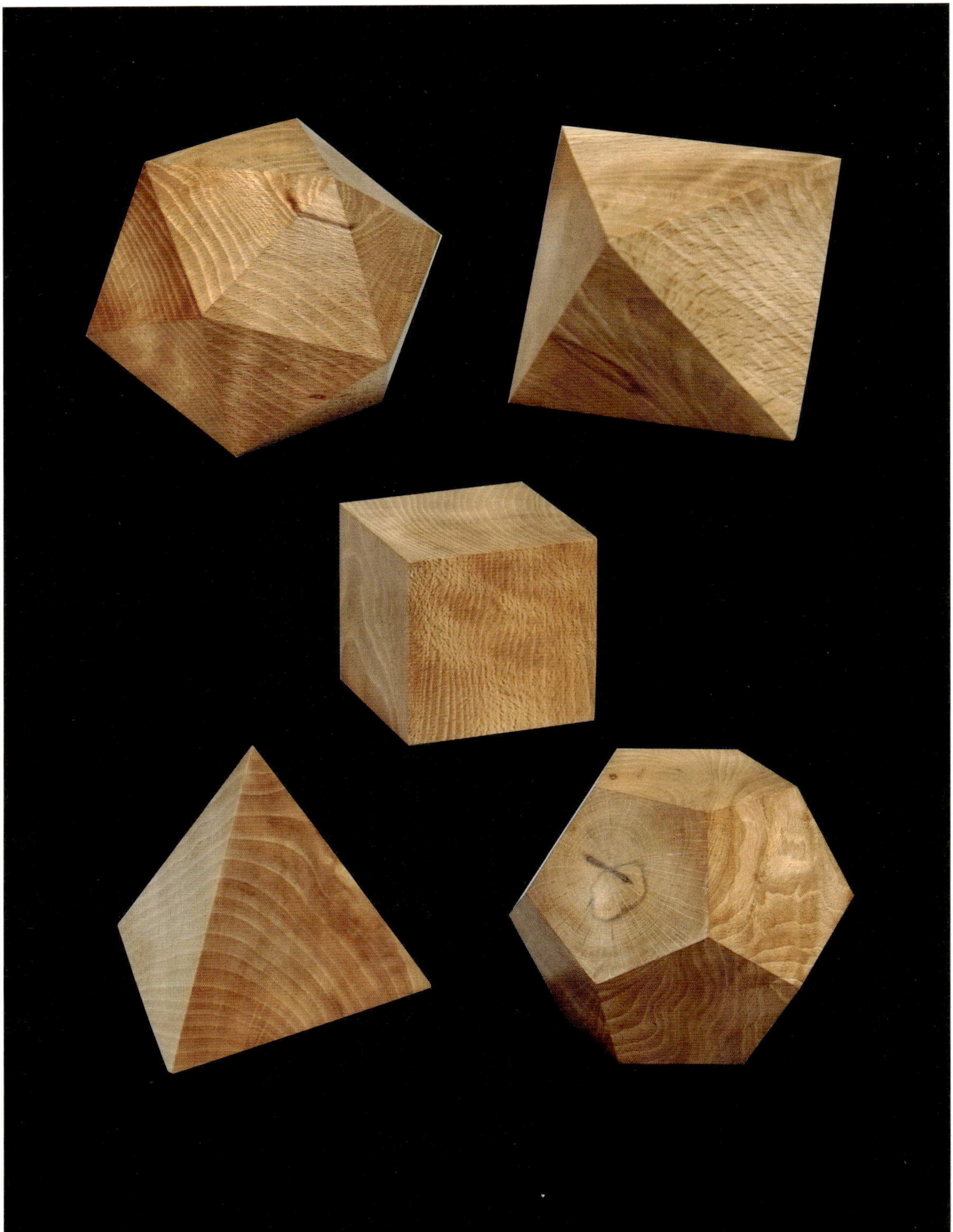

Wooden Platonic Solids,
Martin Turner, 2016

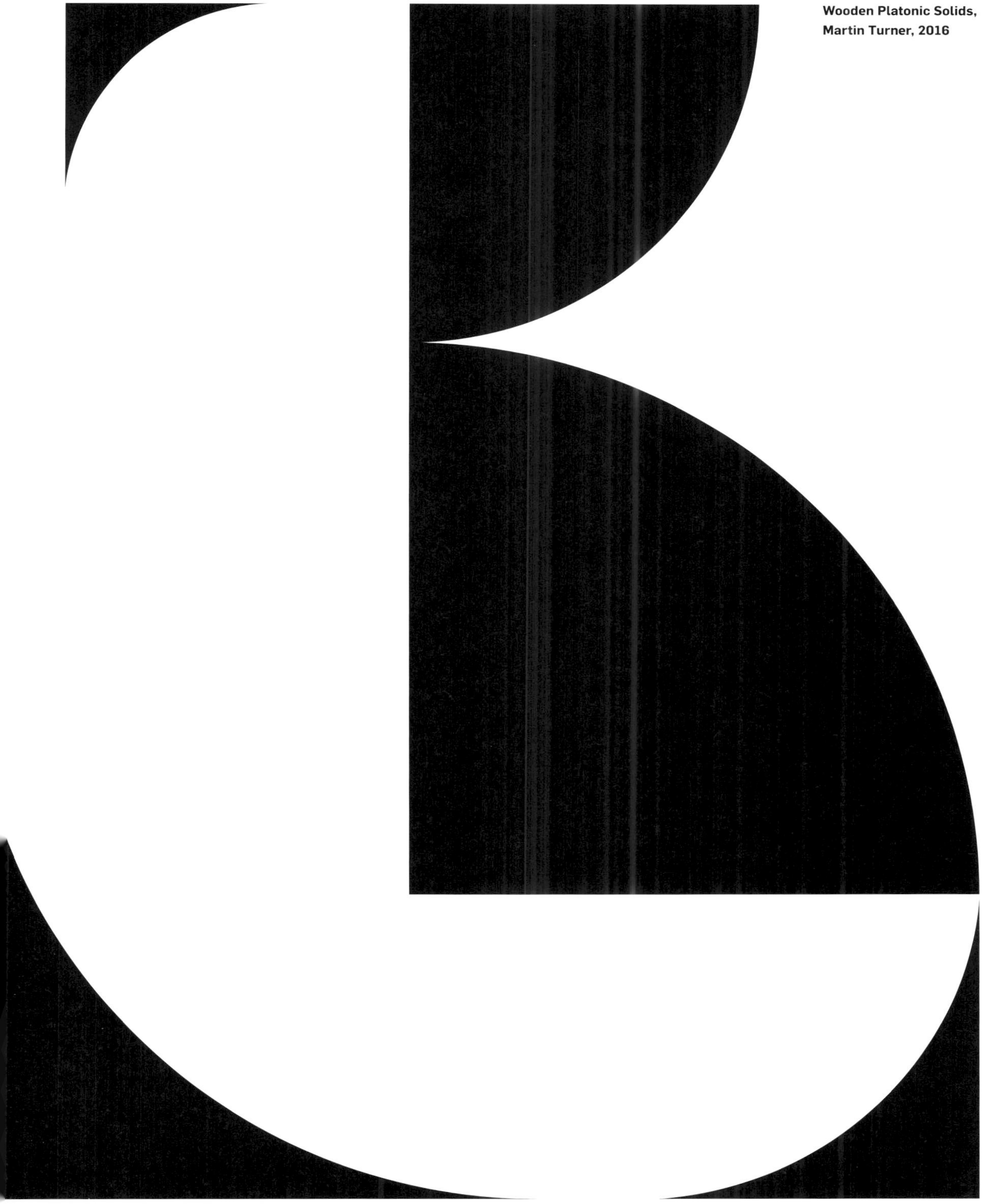

Aesthetical arrangements
Mediation of meaning in Caspar David Friedrich & Adolph Menzel

Werner Busch

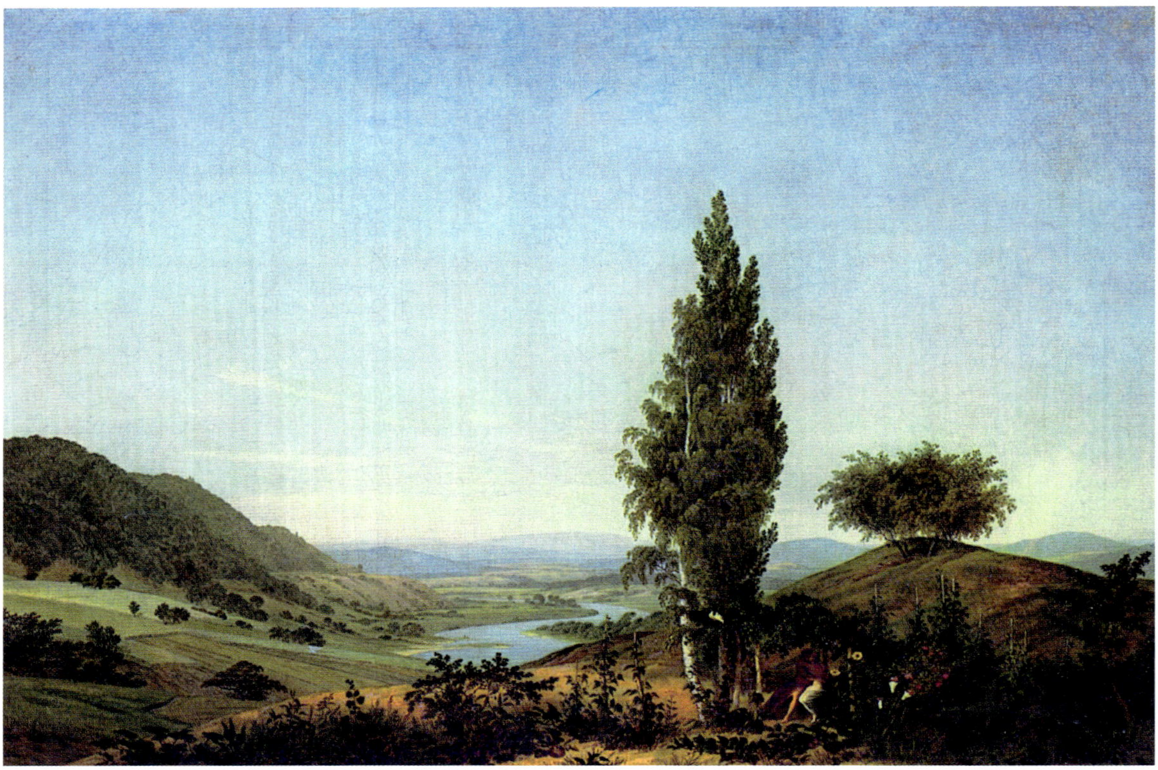

Figure 1. Caspar David Friedrich, Summer, 1807, oil on canvas, 71.4 x 103.6 cm, Bavarian State Painting Collections, New Picture Gallery, Munich

Beware! In art history, it has been claimed that the golden ratio underlies the aesthetic structure of innumerable art works from the classical era to today. This may undoubtedly be true in the one or other case. But scholarship cannot say where the particular artist is supposed to have gained the knowledge of this proportional division found to be so aesthetically pleasing.

Neither theoretical treatises on art in the classical tradition nor any instruction manuals for artists actually mention the golden ratio at all. Certainly, some artists would have read Euclid's *Elements*, and in particular Book II, Proposition 11 with its definition of the golden ratio which this great Greek mathematician came across while investigating the properties of a regular pentagon. All the diagonals drawn inside a regular pentagon form a star shape, and all the points of intersections along those diagonals divide the lines into major and minor segments in the golden ratio, i.e. the ratio of the entire line to the larger divided segment is the same as the ratio of the larger segment to the smaller one.

Aside from Euclid, Luca Pacioli's *De divina proportione*, published in 1509, is the real *locus classicus* of the golden ratio. Moreover, in Pacioli's case, there is evidence of contact to artists. It may well have seemed appropriate to employ a proportion attested as divine in an era when art was liberating itself from its craft bonds, and a theoretical superstructure was being propagated which ascribed creative qualities to the individual artist. Yet here too, the evidence is more than sparse. It was only very late, as the nineteenth century was drawing to a close, that artists were actively encouraged to apply the golden ratio. And the first documented, conscious use of the golden ratio in the artistic process came even later, in the twentieth century, for instance in Le Corbusier's architecture.

Given this background, claims put forward in the 1990s that artists had intentionally used the golden ratio before the end of the nineteenth century were regarded as intrinsically suspect, and viewed as the ineradicable *idée fixe* of art historians obsessed with form and structure. Nonetheless, this basic reservation ought to be taken seriously, and a deliberate application of the golden ratio only assumed before the late 1800s if the evidence is genuinely incontrovertible and the use of this system not only has consequences for the picture's formal composition, but also for its meaning. Here, it is hardly enough to simply point out that, following our European habit of reading from left to right, the ›pictorial protagonist‹ is often shifted slightly off centre to the right, and hence set close to the golden ratio's right perpendicular. In fact, we do read this point as the composition's intellectual and spiritual centre. A main figure set exactly in the centre of the composition would be perceived as immobilising the action. In compositions of axial symmetry, the abstract structure of the pictorial surface dominates, dissolving movement and triggering a less pleasant feeling. It seems, then, as if there is pan-European agreement on where a pictorial protagonist is best placed in a composition – although, admittedly, that position is not necessarily connected to the relations expressed by the golden ratio and is usually an approximate congruence derived from painterly conventions.

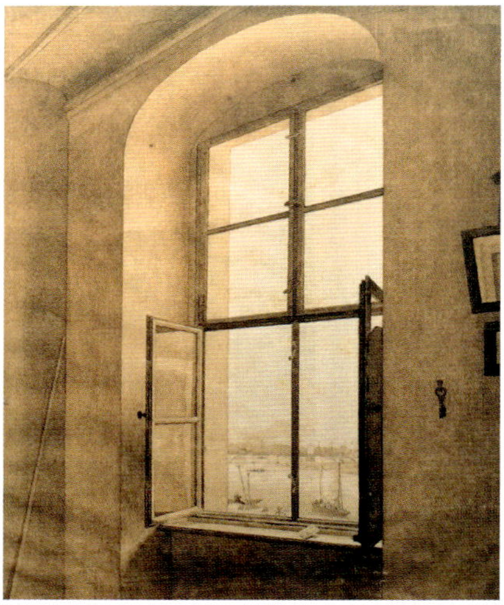

Figure 2. Caspar David Friedrich, View from the Artist's Studio, Window on the Left, c. 1805/06, sepia, 31.4 x 23.5 cm, Belvedere, Vienna

However, such a convention cannot be the issue in the work of the two artists discussed below. Neither Caspar David Friedrich nor Adolph Menzel could be considered as restricting themselves to a compositional structure of the protagonist placed at the location described above. Instead, they not only created a multiplicity of spatial divisions following the principle of the golden ratio, but such divisions, above all, are accurate to the millimetre. This is evident, first and foremost, from the way the vertical and horizontal lines of the golden ratio intersect in many of their works, marking a very particular point in the picture. Moreover, once the viewer has realised where this point is, it proves to be a key element in mediating the meaning of the work.

Caspar David Friedrich

After studying at the fine arts academy in Copenhagen from 1794 to 1798, Caspar David Friedrich settled in Dresden. There he earned a living primarily from his large, pictorial sepia landscapes of popular features on the island of Rügen off the German Baltic coast, where he spent some time sketching in 1801. These sepia works proved very successful, but Friedrich had set his sights on a career in oil painting. Directly before he made the transition to oils, he composed two sepia companion pieces of the windows and view from his studio. These works, which date from 1805/6, were quite clearly painted with a particular agenda in mind and showcased the principles he followed in his art.

The view of the window on the left, seen at an angle (Fig. 2), arranges the objects in the composition so that all of the golden ratio's four lines, two perpendicular and two horizontal, come into play. The left perpendicular marks the angle where the window meets the scuncheon, the inner part of the frame. The upper horizontal meets the upper frame of a cropped mirror to the far right; in contrast, the lower horizontal is marked by the tiny head of a nail in the wall next to the window and which, surprisingly, has a key hung on it. Initially, the right perpendicular seems to run without any specific marker, yet then one discovers that it passes directly through the centre of a small letter lying on the windowsill. The address on the letter is inscribed in minute script, only legible on the original painting: »Den Herrn C.D. Friedrich in Dresden vor dem Pirnaschen Tor«. This not only describes the location of Friedrich's apartment (the Pirnasche Tor was one of the city gates in old city wall), but also brings the outside world into the studio.

The sepia drawing of the right-hand window (Fig. 3) is executed with a similar structure, which not only underscores the intentionality of the process, but also goes a step further. Here, the left perpendicular leads through the centre of the left wing of the window.

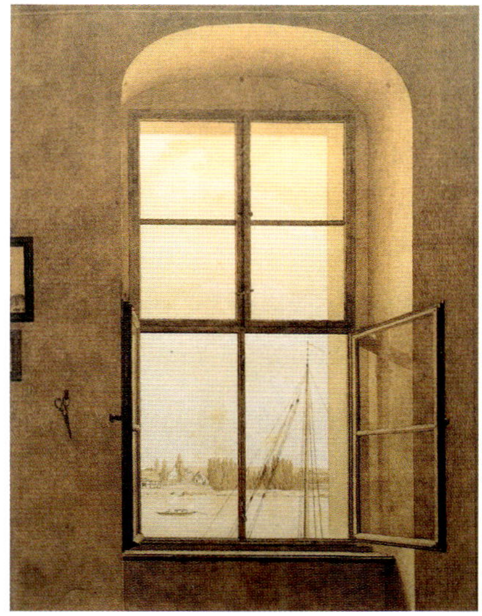

Figure 3. Caspar David Friedrich, View from the Artist's Studio, Window on the Right, c. 1805/06, sepia, 31.2 x 23.7 cm, Belvedere, Vienna

The right perpendicular appears unspecific once again, but one quickly realises that it is aligned with the high mast of a sailing ship anchored in the River Elbe directly in front of Friedrich's studio. In this way, the outer and inner worlds are also formally interrelated here too. The upper horizontal again follows the upper frame of the heavily cropped mirror, here logically located to the far left and solely reflecting the upper part of the artist's face. The lower horizontal line is hardest to identify, since it again leads through the head of a nail, this time holding a pair of scissors. The scissors and key are clearly mutually referential.

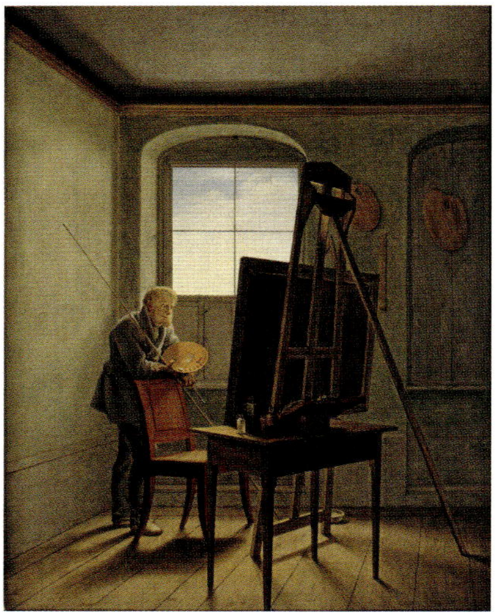

Figure 4. Georg Friedrich Kersting, Caspar David Friedrich in his Studio, Second version, 1812, oil on canvas, 53.4 x 40.9 cm, Berlin State Museums, National Gallery

These two sepia drawings have been the subject of considerable speculation. The results, though, have remained unsatisfactory since the use of the golden ratio was not identified and thus could not be read as mediating meaning. In the right-hand sepia drawing, apart from localisation by the golden ratio, Friedrich also anchors this work with the sheet's central vertical and horizontal axes running through the centre of the cross window.

To put it briefly, the highlighted key might well stand for opening one's eyes to the world, and at the same time for shutting it out. In Friedrich's view, artists had to appropriate existing reality and then, in the artistic process, close themselves off from it to focus entirely on their inspiration without fear of being distracted. He also expressed this idea in his writings, noting on the one hand that, »You must see with your own eye, and faithfully reproduce how things appear to you...«, yet on the other, »In hours of inspiration ... [art]... becomes intuitive form«. However, Friedrich added, there was a necessary prerequisite for this: »Close your bodily eye, so that you may see your picture first with your spiritual eye. Then bring to the light of day that which you have seen in the darkness so that it may react upon others from the outside inwards.« For Friedrich, the relationship between inner and outer, sentiment and reality, spirit and material was central, and this may be exactly what the key symbolises. The pair of scissors might then express the fragmented nature of perception, yet at the same time testify to the divine principle of a ›golden ratio‹ ordering the fragmentary experience of reality. In this way, the work addresses the issue of the relationship of the part and the whole, as well as fragmentation resolved by an abstract order which has a cosmological dimension of its own. The mirror reflecting the artist's eye and forehead could well be read metaphorically for the process of transforming the seen in the mind of the artist. Certainly, there was a good reason why, for centuries, the eyes were regarded as the mirror of the soul. In that sense, the left-hand sepia drawing could reference the bustling life in the real world, while the right-hand work evokes reality's metamorphosis in the artistic process through absolute principles of order.

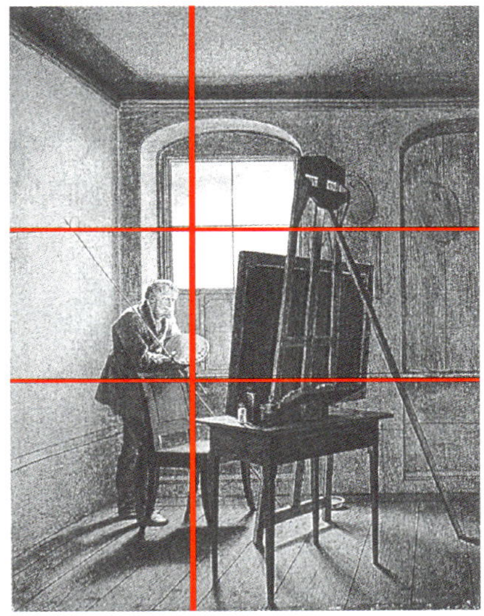

Here, a portrait by Georg Friedrich Kersting showing Friedrich in front of his easel in his studio may well provide evidence that such an interpretation seems to be moving in the right direction. Beforehand, though, let us briefly turn to Friedrich's *Summer* (Fig. 1, p. 72), a landscape from 1807, which would appear to be the first example of the golden ratio in his oil paintings. In the landscape, which by Friedrich's standards is exceptionally serene, a couple is cuddling up in a bower with a birch and poplar to the left. The two trees are rooted close together, their foliage intertwined and entangled – so much so that they appear to form one single treetop. The line through this treetop marks the central axis of the two trees, which come complete with two white doves billing and cooing among their branches lower down, echoing the loving couple and personifying the happiness of summer. Nothing else in this work projects as far above the horizon as these two trees busily engaged in becoming one, and their axis forms the right-hand perpendicular of the golden ratio, exactly to the very millimetre. Here, then, this line represents the all-embracing harmony between human beings and the natural world. In the companion piece of a winter landscape, it transpires that happiness does not last, but that work is only incidental to this discussion.

Kersting's second version of a studio interior with Friedrich standing in front of the easel dates from 1812 (Fig. 4, p. 76). Here, the shutters of the right-hand window are closed entirely, just as recommended in manuals for artists. Since only the lower shutters of the left-hand window are closed, the light falls into the room from just one source. This was the favoured lighting for painters, with light preferably from the north. Friedrich can no longer look out into nature; the natural world is shut out, leaving nothing to interfere with his artistic inspiration – one could say, his bodily eye is closed. Friedrich is holding a mahlstick and palette in one hand, with the brush in the other, visible below the palette. The process of painting is about to start. In this work, rather than a mirror on the wall between the windows, there is another palette set above a ruler and set square, each hanging on a nail at the same height. And as we can already guess, the top of the set square and the ruler lie exactly on the golden ratio's upper horizontal line. The lower horizontal passes precisely through the point where Friedrich's finger meets the brush, and the line continues along the sill of the closed right-hand window where a slim book, hardly visible, is lying on the sill. This book only gains its significance when one knows that the golden ratio's left perpendicular also passes through the point where the finger and brush meet – highlighting, in other words, the divinely inspired starting point of the subsequent painting process. Since this compositional structure directly connects the origin of the creative process and the little book on the right-hand windowsill, the book is then also endowed with meaning. This can, thus, only be Friedrich's sketch book, where he faithfully rendered the natural world and reality. The sketch book supplies the material which, through the artist's inspiration and with the help of a harmonious system, results in the picture.

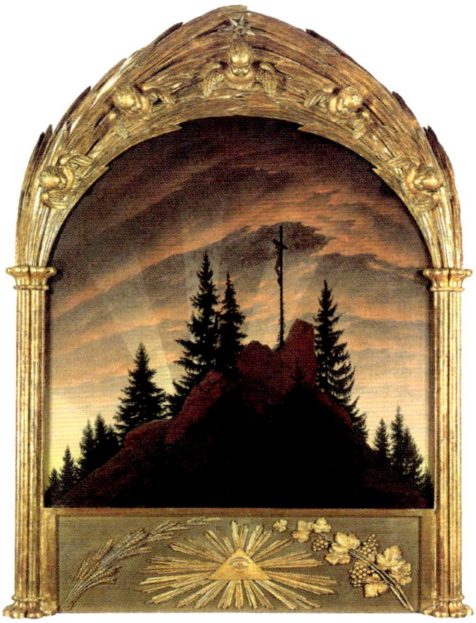

Figure 5. Caspar David Friedrich, Cross in the Mountains (Tetschen Altar), 1807/08, oil on canvas, 115 x 110.5 cm, Staatliche Kunstsammlungen, Galerie Neue Meister, Dresden

From then on, the golden ratio became Friedrich's main structural principle, supplemented by such geometrical figures as conic sections, hyperbola, parabola and ellipses. Here, though, we only have space to briefly consider Friedrich's most famous work, the Tetschen Altar (Fig. 5), completed in late 1808.

In this painting, all four lines of the golden ratio create an unmistakably pointed effect. Inevitably, the path of the golden ratio's key line, the right-hand perpendicular, passes through the vertical beam of the crucifix on the mountain. The left-hand perpendicular, in contrast, runs through the axis of the highest pine tree as well as through the central ray of the five broad bands of light radiating from the sun setting behind the mountain. If one connects the rays, they lead to the base of the altarpiece meeting at the exact point where the golden ratio's left-hand perpendicular touches the base. The lower horizontal line leads precisely to the foot of the cross where it intersects with the right-hand perpendicular. This structure particularly accentuates the point where the cross rises from the mountain. Evidently, it is a structure inspired by the words of Saint Paul »petra erat Christus«, Christ as the rock of faith. In relation to the organic compositional structure, the upper horizontal line appears to run without a specific purpose until one realises that it rests directly on the projecting column imposts of the carved gilded frame, which was also designed by Friedrich. From this point, the palm fronds lean towards the centre, forming a gothic shape to frame the picture. Five winged putti are set in the frame, each crowned by the evening star, which is also the morning star. In number, the five putti correspond to the five broad rays of the setting sun. The base is decorated with the signs for the bread and wine of the Eucharist. At the same time, the structure of the frame is analogous to a cross-section through a nave. When the Church of St. Nicholas in Leipzig was refurbished in the later eighteenth century, the nave columns were decorated with palm fronds opening upwards from the imposts to the gothic vaulted roof. When one considers that in ecclesiastical thought the congregation's worldly space extends to the height of the column capitals, and above that is divine space, then Friedrich's upper horizontal of the golden ratio also marks the transition from the earthly to the heavenly sphere. Moreover, at the same time it aesthetically connects the space of the painting's real landscape with the Christian signs inscribed on the frame – which, moreover, could only be related in this way.

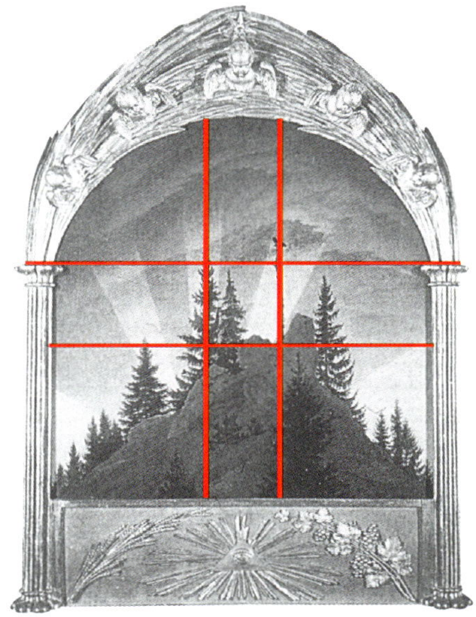

Adolph Menzel

In the case of Adolph Menzel, the discussion can be somewhat shorter since in principle Menzel used the golden ratio in a similar way – as a method of mediating meaning. However, Menzel's work lacks the hope of the golden ratio's ability to resolve aesthetically the experience of reality disintegrating into perceptual fragments in a process that may even offer comfort in transcendental experience. This might have been conceivable for the Romantics, although in Caspar David Friedrich's case one should also note that, as a staunch Protestant, the hope of eternal life only came from the journey through death. For him, there was no certainty in life. In contrast, Menzel accepted the chaos of present reality as a given. His aim was to make life, just as it is, bearable and, and with the help of his art, constantly sought to shape it into an agreeable form – and here, he enlisted the structural principle of the golden ratio to help him. He set up an aesthetic system to confront the reality besetting him, yet did so without ever opening up the possibility of there being a permanent dialectical resolution to this conflict.

Here, we consider two examples of Menzel's programme for coping with life, one from an early phase and one from the late phase of his life. Soon after Menzel's death, two books were published, one by Julius Meier-Graefe and the other by Karl Scheffler, celebrating his oil sketches from the 1840s as the sole highpoint of his oeuvre. Here, it was said, Menzel painted freely, nothing less than proto-impressionist, working on the level of the French artists. Everything that followed, the argument went, was entangled in detail, laboured, ultimately conventional, including Menzel's paintings of events from Friedrich the Great's reign; these were nothing more than reconstructed costume pieces, born out of an obsolete notion of history painting. This image of Menzel's art created in the early twentieth century proved to have a long-lasting effect. His oil sketches were much-loved, while the rest of his works, by far the larger part and including his late crowd scenes inspired by bustling urban life, were at best respected for the effort he put into them.

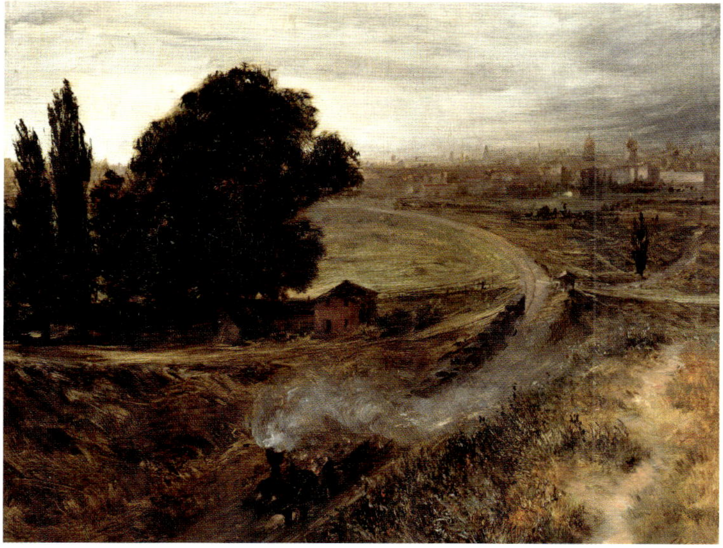

Figure 6. Adolph Menzel, The Berlin-Potsdam Railway, 1847, oil on pasteboard, 42 x 52 cm, Berlin State Museums, National Gallery

Today, it is evident that this reading of Menzel's oeuvre distorted both the view of his oil sketches as well as his later works. Among his oil sketches, along with his *Room with a Balcony*, the *Berlin-Potsdam Railway* (Fig. 6) was his most admired work. In his sketches, he was praised for his relaxed and light painting style. Moreover, it was said, in his *Berlin-Potsdam Railway*, Menzel had appropriately captured the pace of the train rushing towards us – lending a fleeting appearance a vivid and graphic form. Such a view was both right and wrong. The painting style is indeed light and fluent, and the composition breathes vitality and a perfectly harmonious mood. But we need to remember that this approach was deliberately chosen, and not directly reproducing the experience in front of the painting's subject. The Berlin-Potsdam Railway from 1847 is based on a preliminary sketch from 1845. The latter is considerable more precise than the subsequent oil sketch, quite evidently produced in the studio. And since no attention was paid to the consequences this had, it was never realised that this painting has a precise underlying structure.

What appears so picturesque here depicts a city outskirts, a zone of transition, neither fish nor fowl. The line of the track for this new means of transport cuts through this disorderly transitional zone, following an exact curve. The line from Berlin to Potsdam started at the Anhalter Bahnhof terminus station. Since that station was only built in 1841, the experience portrayed here was new. Moreover, since the railways, an invention imported from England, ran at a frenetic pace, it was also a bewildering and confusing experience. So this work shows two things which people at that time had to come to terms with: first, the disorderly transitional zone of a city rapidly expanding beyond its walls and, second, the new, powerful and, quite literally, incisive system of burgeoning industrialisation.

Menzel turns to the golden ratio to try and make this experience tolerable. The engine's chimney, the epitome of power released, lies exactly on the golden ratio's left-hand perpendicular. In the preliminary sketch, the curve of the railway as it rushes out of the city did not have, as here, a branch line at the height of the little gatekeeper's cottage – but Menzel's decision to include it emphasises the importance he attached to it in the organic composition. In the upper and lower sections respectively, this parabola shape marks the upper and lower horizontal lines of the golden ratio, with the gatekeeper's cottage as the crown of the arch – and this cottage, as odd as it sounds, is set at the intersection of the golden ratio's construction lines. Were this emphasis on the compositional abstract framework not found in other works by Menzel as well, such an argument might sound like projection. However, when one additionally notes here that the edge of the central cottage's right wall, from our perspective, is set exactly on the axis of the compositional centre, which is further emphasised by the edge of the foliage on the large tree, one realises that Menzel's free painting style follows a predefined, virtually absolute system. Art may compensate for the new, unfamiliar experiences, yet it does so without neutralising them.

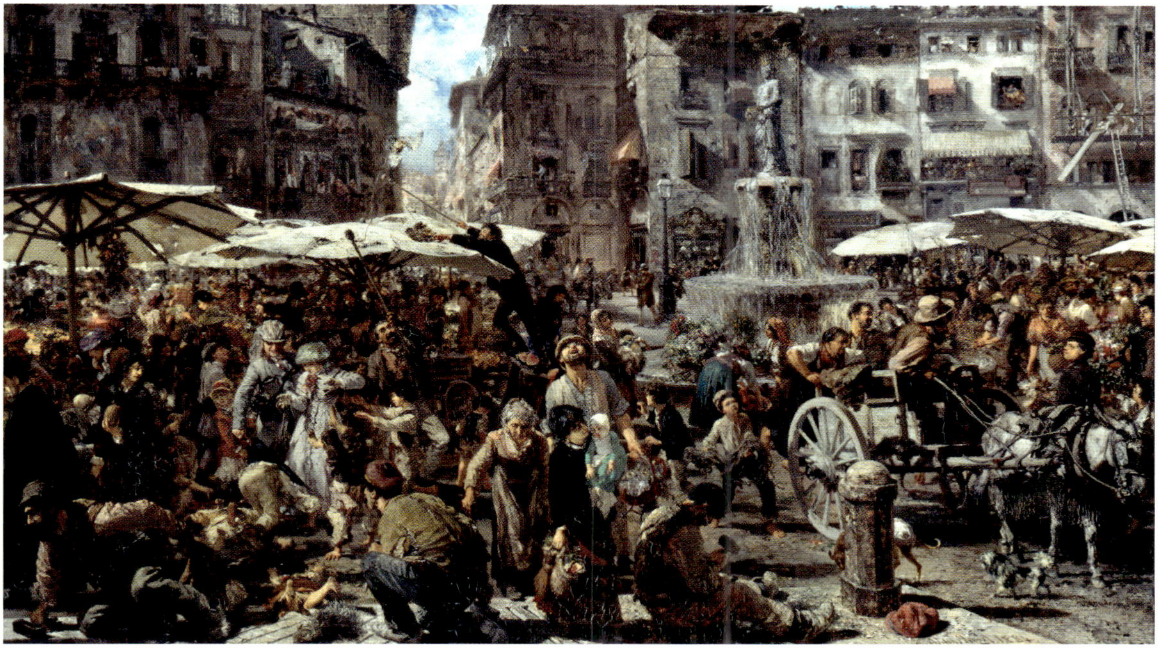

Figure 7. Adolph Menzel, Piazza d'Erbe in Verona, 1882–1884, oil on canvas, 74 x 127 cm, Staatliche Kunstsammlungen, Galerie Neue Meister, Dresden

The second example is Menzel's *Piazza d'Erbe in Verona* (Fig. 7, p. 83), his last large format oil painting. In this work, which dates from 1882 to 1884, the beholder inevitably has an impression of overall chaos. A teeming crowd of people, packed tightly together, has gathered on the market place around the Madonna Verona fountain. The innumerable market stalls are protected from the sun by identical white sunshades, but the stalls themselves are virtually lost in the surging masses. Indeed, without the title of the work and the fountain, as well as the knowledge that Menzel visited the town three times making approximately 100 detail studies, it would be hard to know for certain that this is actually the market place of Verona. Moreover, Menzel has included none of the well-known palazzi lining the square, and only depicted a collection of buildings typical of the area. A surprising multitude of people are also thronging the many balconies facing the square, as if they had previously arranged to gather at the same time to watch events in the market below.

There is also more than enough hustle and bustle on the market square itself. To the left, an English couple, dressed in light blue clothes and accompanied by their daughter and her governess, are warding off a hoard of Verona's street urchins. In the foreground, workers are laying paving stones and, to the right, a donkey cart is battling its way through the dense crowd. Building work is also underway on one of the houses in the background to the right. There is movement and change wherever you look. Slightly offset to the left of centre, a street leading to the market opens up a vista to a tower which, in reality, is not visible from the square. A patch of blue sky rises above the tower – and the tower itself, as a visual focus, also marks the left perpendicular of the golden ratio. But for the work's expressive character, this line has a far more important function. It leads to the point where, with his hand reaching as far as he can, a bird-dealer is desperately trying to catch one of his birds tied to a string in order to take it down from the large sunshade where it seems to have snagged its claws. The bird-dealer is straining forwards, wobbling on a shelf already at a dangerous angle.

At any moment, it will tip over, triggering a totally chaotic chain reaction through the entire market. The role of this point as the secret centre of the overall composition is further suggested by the fact that, at the same time, the golden ratio's upper horizontal line passes through the hand and bird. The vertical and perpendicular lines cross precisely at this culmination point. Moreover, the structure of the composition directs our gaze exactly here. Since the horizon line is identical with the golden ratio's upper horizontal – and Caspar David Friedrich often combined these two lines as well –, this is the vanishing point of the entire composition.

On the one hand, then, Menzel has applied a system of order which, amazingly, marks the starting point of total chaos. On the other, and rather unusually for him, in the face of the absolute pandemonium at the market, he apparently felt the need to integrate a kind of triangular gable in the foreground set around the axis of the centre of the composition. The highest point of this triangle is the head of the water seller as he looks up. Since for this view, Menzel would be looking out from a first-floor balcony facing the market, and as the beholders we are in the same position, the water seller seems to be offering his goods to the artist and, hence, to us. Two workers form the sides of the triangle, one is absorbed by the altercation with the English couple. Two women, an old lady with a stick, and a young woman with a child on her arm, are approaching us, laden with their shopping from the market. The young boy, again on the axis of the compositional centre, appears to be nothing short of falling towards us, and so opens up the picture to the front. He marks the aesthetic border to the beholder. This triangular gable, an unmistakable structural element, is designed to give a certain stability to these chaotic events, although here too, this device does not neutralise the confusion.

So while Friedrich dreams of salvation with the help of the golden ratio, Menzel appears to no longer believe it possible. Instead, he presents the experience of modern urban life, something he first came across in Paris. If he here forcefully inscribes a system of order onto it, this only represents the ultimately futile attempt of standing one's ground against something so overpowering.

Busch, Werner: Caspar David Friedrich. Ästhetik und Religion, Munich, C. H. Beck 2006.
Busch, Werner: Adolph Menzel. Auf der Suche nach der Wirklichkeit, Munich, C. H. Beck 2015.
Fredel, Jürgen: Maßästhetische Studien zu Proportionsfragen und zum Goldenen Schnitt, Hamburg, Lit 1998.
Hemenway, Priya: Der geheime Code. Die rätselhafte Formel, die Kunst, Natur und Wissenschaft bestimmt, Cologne, Taschen 2008.
Corbalán, Fernando: Der Goldene Schnitt. Die mathematische Sprache der Schönheit, Cologne, Librero 2016.

»To see is only to touch more accurately!« The Church of St. Canisius

Heike Büttner

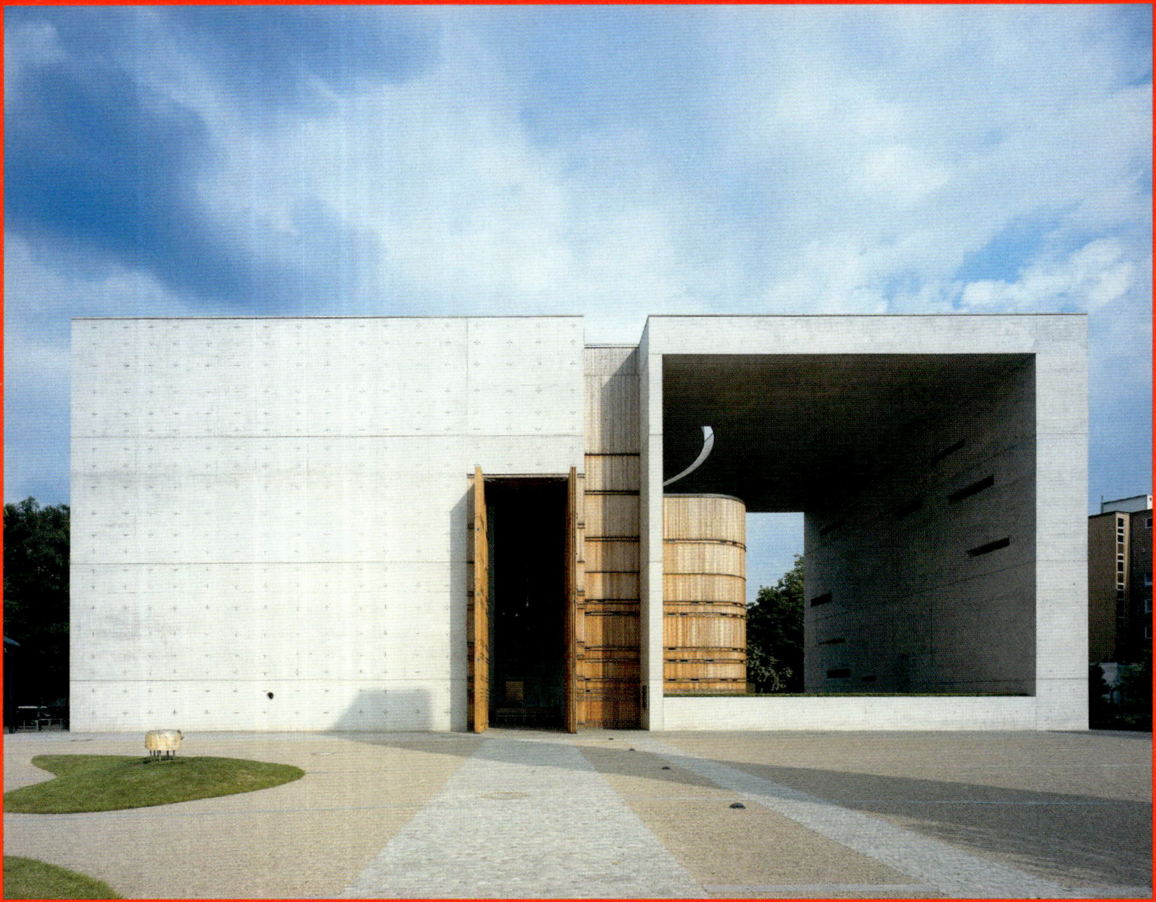

A square surrounded by the silhouettes of fire break walls, pink-coloured walls, bundles of cables, a destroyed identity, a concrete wall with a cross and the body of Christ from a crucifix which had survived. To begin with, the architect always has to create something after some kind of destruction. In our memories, meanings may perhaps shift and change, but it is the signs which carry a message for us that determine the design and dissolve any arbitrariness. Time and again, individual observations evoke a different design, in turn a memory of something different. If a space can become a sign, can it additionally signify a thought? As architect Louis Kahn once said, »To see is only to touch more accurately!«

Mass, transparency, landscape

The Church of St. Canisius (consecrated in 2002; Fig. 1) is set at the empty heart of a perimeter block development in Berlin close to Lietzensee lake. The expansive size, the near unified quadratic yet simultaneously rectangular body, the foundations set off from the ground at a height of 160 cm, all lend the church a material sublimity. The dimensions of the openings and recesses form a relationship with their density and highlight the design of the church as a bridge construction. Its spatial configuration comprises an enclosed and an open space. The architectural design exhibits a complementary figuration of empty and full, a painterly game of mass and transparency. The articulation of the adjacent structural components to the landscape and the sky make the dialogue of lightness and gravity come alive.

In striving for simplicity, reduction and unequivocalness in the choice of symbolic elements, the church seeks to connect with modernity. In its relations, the church is cogent, deriving itself from self-referentiality. It is dynamic, sensory, accessible in a new way from each standpoint and only offers indirect insights into the interior liturgical space. The diverse diameters of the bodies lead the gaze in different ways. What is created in the eye is a volume with internal tensions and combinations.

Figure 1. St. Canisius in the Berlin borough of Charlottenburg-Wilmersdorf

When we look at the interplay of surfaces and bodies of the open space's ›Angel Wall‹ (*Engelwand*), we realise that the base surface areas of the illuminants have been deliberately cropped to reverse the perspective of the view. Almost as if in a photograph, relationships are created with the surroundings and the hidden life behind the protruding fire break walls.

One space and many places

The experience of being surrounded by space and materiality is already evident as we arrive at the eleven-metre-high portal and in St. Mary's Chapel (Fig. 2; Fig. 3, p. 90). On entering the building, we disturb the balance of a meticulously structured geometry. At this moment, architecture is only an element which engages with us passively. We are penetrating a space governed by principles carefully constructed from architectural and theological ideas. The line can become space and then turn into a line again – the relationships gain in clarity through intersections or divisions and unities.

The room for the church is designed as an empty space with the altar as the centre of the congregation. This room is characterised by a longitudinal and transverse orientation, taking up the idea of the early basilica. The geometry is informed by circular and rectangular shapes which resonate into the open space. The nave's central element is an incised circle, whereby one segment of this circle is absent. The ground plan reveals the diversity of the interplay between the various spatial points. The figures are intertwined: a circular, lower area with the tabernacle and a light shaft to the sky. The liturgical places are simultaneously structural and integrative elements. They are places for gathering together, and set in the open space of the room. The altar and apse, the tabernacle and the Eternal Light, the font, the ambo, the raised stand for the Bible, and the chapel as a centre for Marian worship form the inner core of the church. The apse is aligned with the nave, with the altar for communion and the tabernacle where Christ dwells forming the heart of the congregation.

Recurring proportions, perceptions

In the Church of St. Canisius, the golden ratio has been applied to many different aspects of the design. As a classic law of proportions, the golden ratio is an overarching principle embracing both interior and exterior space, and representing the transitions between the individual sacral locations. It is a visible criterion which can be taken as an aid to compare one element directly with the others. As a result, the individual components are connected to each other, and the relations between them can be established through diversity as well as similarity. Some divisions can be read as sub-divisions, others as combinations. The twelve apostle crosses, which appear fragile and delicate, show that even the smallest external detail already exists on the large scale.

The verticality of the enclosed and open space offered the chance to juxtapose materiality and light with their respective absences. The discontinuities in the outer contours of the church create mistaken attributions; the interior then starts to bring these elements together. Lines, surfaces and space eluding us at the margins of perception and becoming a fiction create an uncertainty of relations. The fall of light in the apse appears to be an upright rectangle behind the boundary of the reredos: the seemingly monochrome light, falling through the opaque, curved glass, is intensified by the two-dimensionality of the altar wall.

Since, depending on the point of view, it is cut at various places, it can hardly be seen as a soaring rectangle, and so we are constantly confronted with a detail and a rectangle which needs to be completed in the mind. Is this a view of an upright rectangle or a portrayal of a column of light in the form of an upright rectangle? In fact, the altar wall and the apse behind it are identical. Is it the wall or the incident light which transforms planarity into spatiality? In the room of the church, we look through a translucent sheet of glass from one area into the other; this is an experimental modern interpretation of the apse behind the altar area. This opaque glass optically underscores the division of this space so that something seems to remain undefined, in suspense, negating the real transparency of glass.

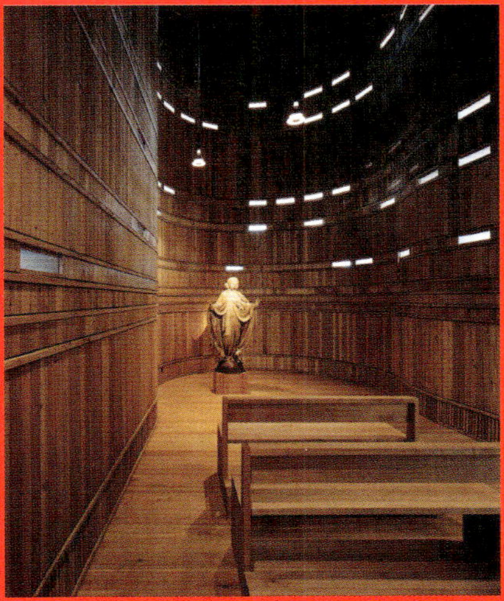

Figure 2. St. Mary's Chapel, centre for the worship of the Virgin Mary

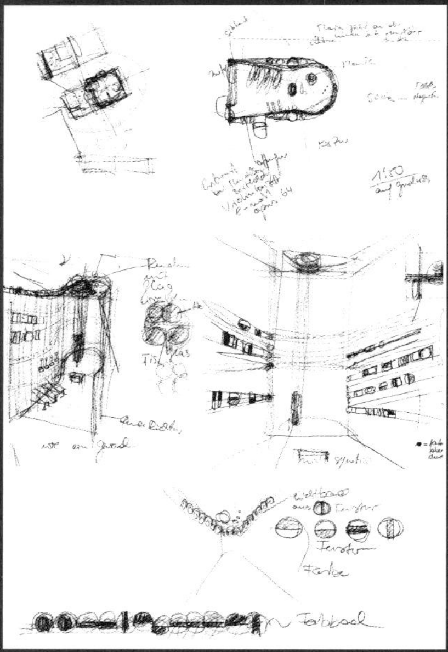

Figure 3. Design sketches for St. Mary's Chapel

Interior, exterior, symbolic incisions

As the southern perimeter, the space opens out as an external raised level with St. Mary's Chapel and the ›Wall of Water‹ (*Wasserwand*). In the vista of the church, the golden rectangle is identifiable as the continuation of St. Mary's Chapel into the open space. The geometry of the golden ratio is tangible in the ground plan, elevation and section; from the smallest element of the concrete inlays as the fasteners on the anchoring cones to the Apostle crosses and the allusion to a circle as a symbolic incision, as a source of light and metaphor. Geometric forms such as squares, rectangles, regular pentagons, cubes, tetrahedrons or cylinders are only intended to be recognisable as allusions (Fig. 4). The large glass surface between interior and exterior serves to dissolve boundaries. It represents a different spatiality, creating a balance between stability and movement, weight and lightness. The open space, the ›heavenly Jerusalem‹ as the place of the creation, is designed as a kind of cupola on its side, yet can also be read as a flanking aisle; limited by being closed off to the sky by an element that includes the geometry of the church emblematically as a segment of a circle. This expands the interior of the church both symbolically and spatially. The dialogue between these two locations is generated through the act of the liturgy. The idea of the open space is rooted in the interior/exterior juxtaposition as the cycle of life. At the Church of St. Canisius, this is the interpretation of the space of creation, the bride, the ›heavenly Jerusalem‹ in which the four elements meet: fire, water, earth and air as the interior space created by God.

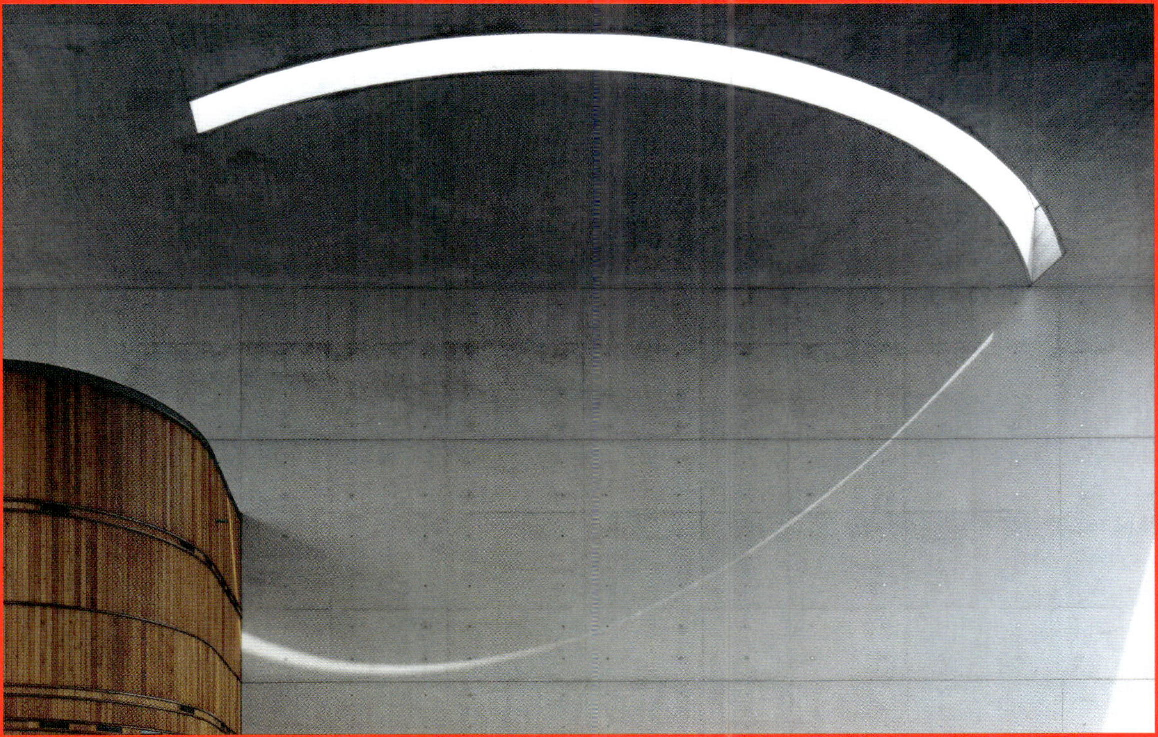

Figure 4. Forms, illuminant and symbolic incision

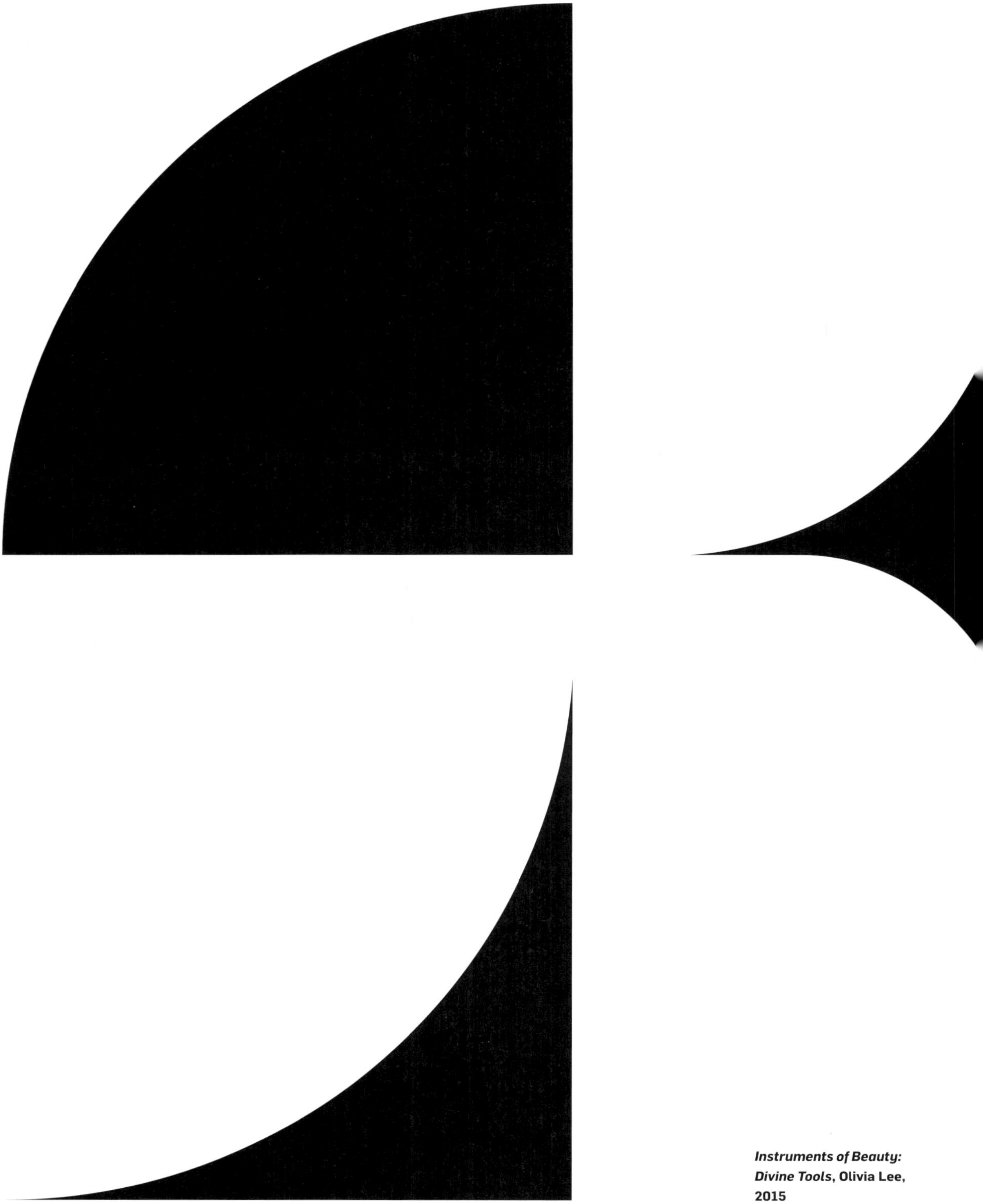

***Instruments of Beauty: Divine Tools*, Olivia Lee, 2015**

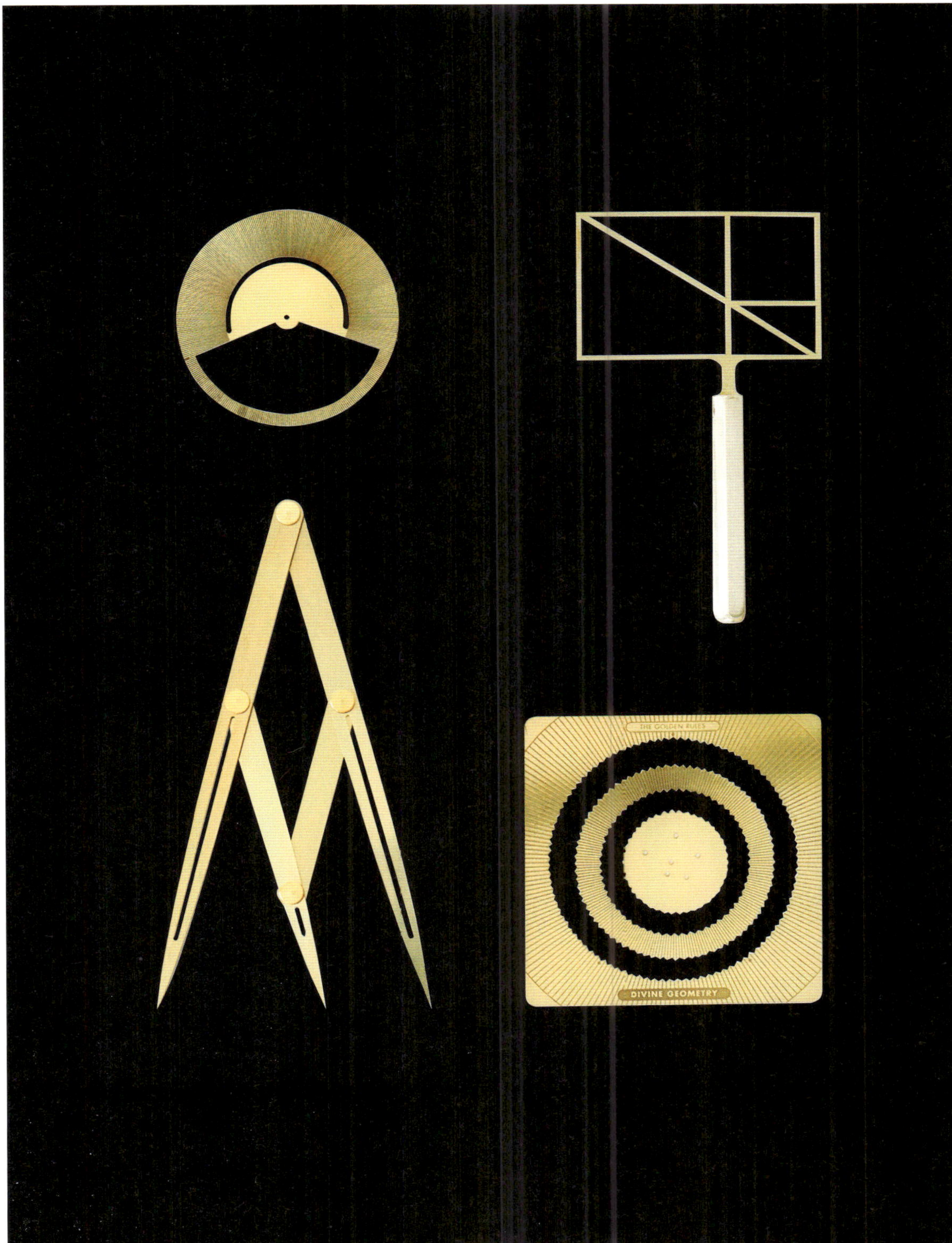
THE GOLDEN RULES
DIVINE GEOMETRY

Grids and rules

The golden ratio in typography and graphic design

Wolfgang Beinert

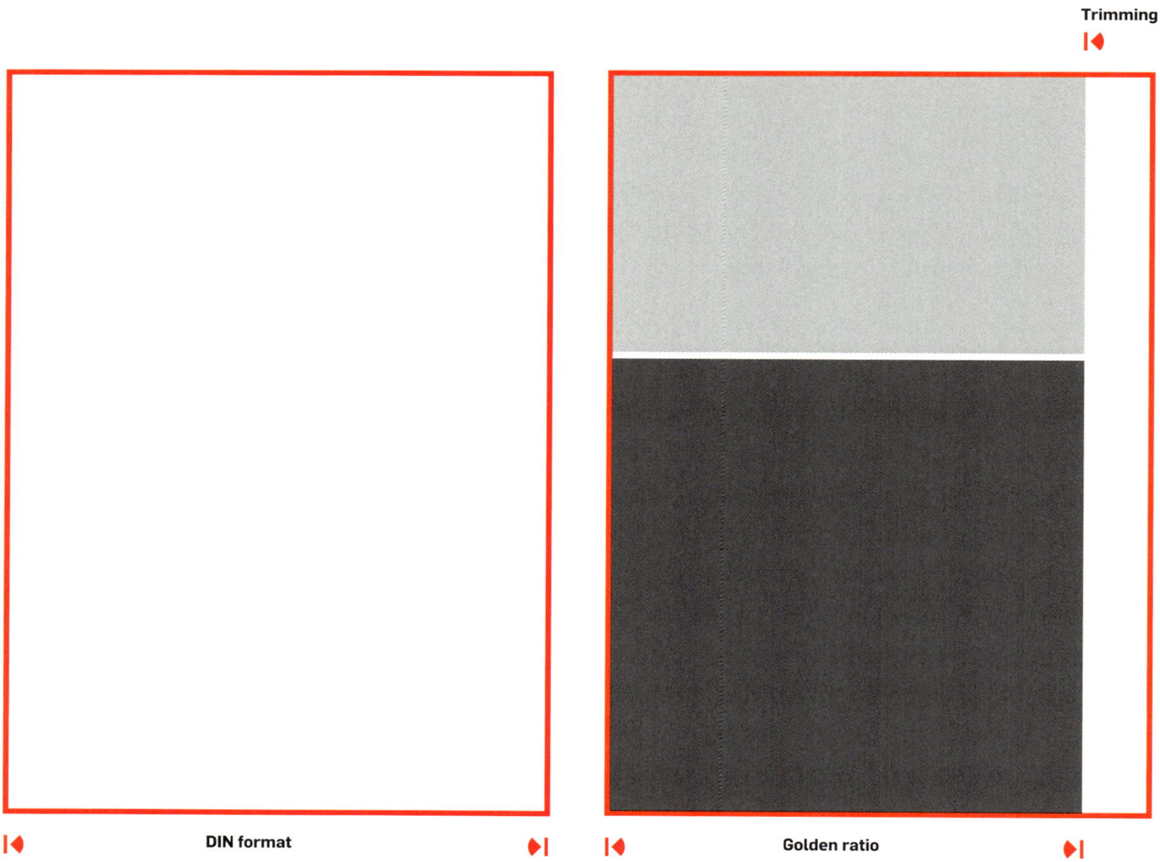

To obtain a ›golden‹ aspect ratio from a piece of standard DIN-sized paper, the paper must be trimmed. The golden ratio is also different from all classic book formats.

In typography and graphic design, there are many ways of applying mathematical ratios and number series to systematically proportion, structure, and standardise formats, layout scaling, and colour grids. This practice follows a long tradition, going back to Johannes Gutenberg (c. 1400–1468) and beyond. Gutenberg, one of the fathers of typography and the inventor of movable-type printing, used an early Gothic canon of page construction known as Villard's diagram to proportion the printed and unprinted areas of the pages in his 42-line bible (approx. 1455). Villard's diagram (like all such canons) was designed to achieve a geometrically exact division of rectangular areas, balanced in shape, proportion, and aesthetics.

The golden ratio – in the strict mathematical sense of the proportion 1:1.618 – can only be used in the graphic design of visual communications for the aesthetic proportioning of a self-contained, two-dimensional area, for example for calculating bleed areas (the area trimmed from a page after it is printed) or dividing a rectangular paper format.

When we talk about the golden ratio in typography or graphic design, we therefore generally mean the Fibonacci numbers 0, 1, 1, 2, 3, 5, 8, 13, 21, 34, 55, 89, 144, 233, 377, 610, etc. Ratios formed from adjacent terms in this series approximate the golden ratio, with each successive ratio being alternately larger or smaller than the golden ratio itself. The ratios are easy to calculate: for any two adjacent terms in the series, divide the larger number by the smaller. Following this rule, we get the ratios 13:8 (=1.6250), 21:13 (=1.6154), 34:21 (=1.6190), 55:34 (=1.6176), etc.

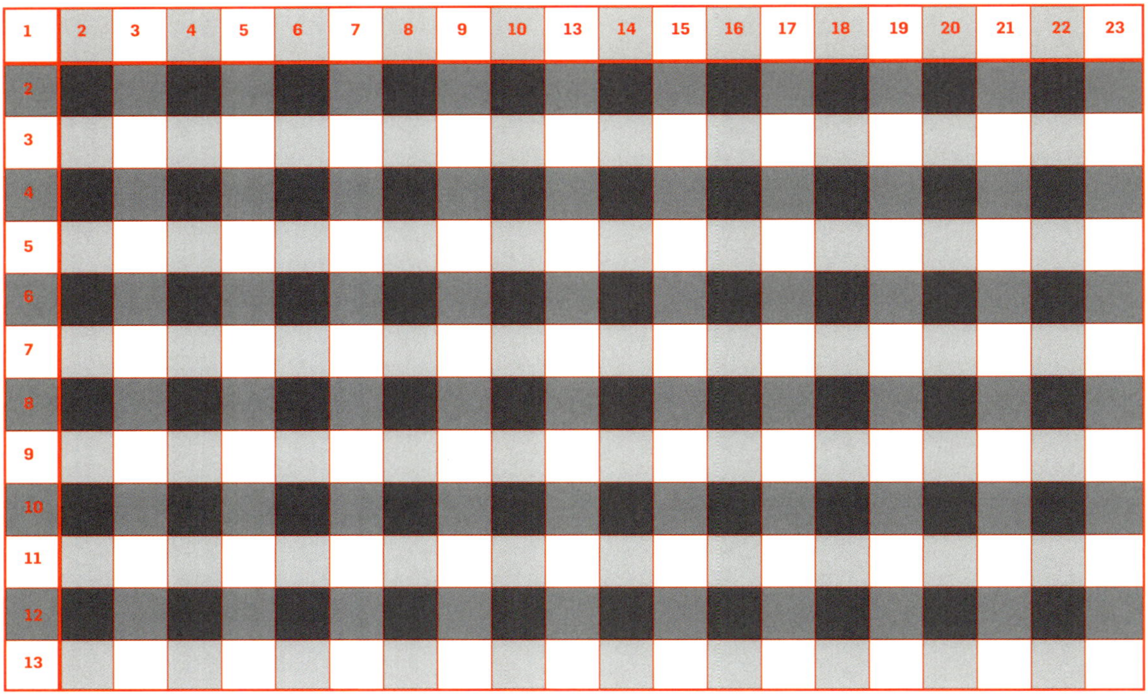

It is possible to create very simple grids based on the Fibonacci numbers, such as this example in a ratio of 13:23. These grids, however, do not meet the requirements of modern layouts and desktop publishing (DTP) programmes.

This number series, commonly used in typology, is named after the Italian mathematician Leonardo Fibonacci (c. 1170–1241), who used it in 1202 to describe the rate of population growth for rabbits. From a mathematical point of view, however, using the Fibonacci series to create a page construction canon, design a typeface, or scale colour and type size on a webpage, has very little and indeed sometimes nothing to do with the proportion of the golden ratio. In these cases, the Fibonacci connection takes the form of a simple, proportionate coordinate system whose smallest square unit of area either corresponds to a term in the series or is systematically linked to it in some other way.

The golden ratio and standardised paper formats

Virtually all paper formats in Germany are defined by the German Institute for Standards (*Deutsches Institut für Normung e.V.*) in DIN standard 476. The aspect ratio of all A-series formats is that of a square's side length to its diagonal, specifically a ratio of 1:√2, or 1:1.414 (rounded). The ratio of width to height is thus approximately 5:7, which deviates substantially from the proportions of the golden ratio.

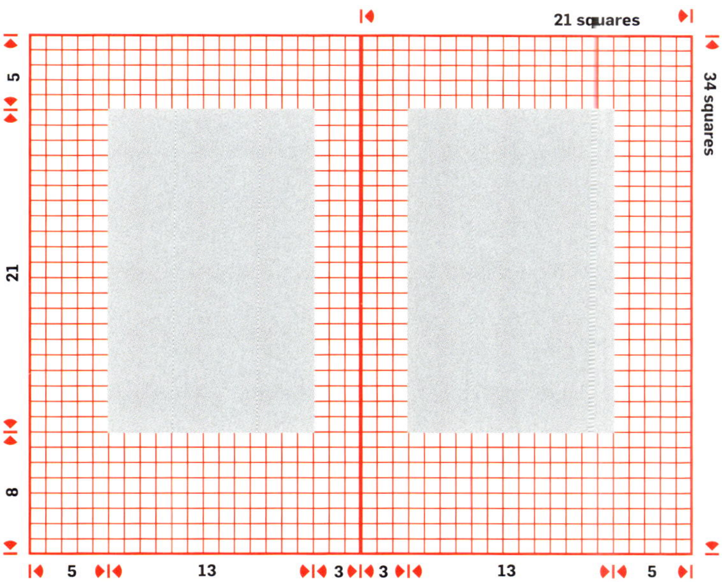

The Fibonacci numbers can be used to create a page format in the golden ratio. The grid, for example, consists of 21 x 34 squares and contains a type area with an inner margin of 3 squares, top and outer margins of 5 squares, and a column of 13 x 21 squares.

Layout grids based on the Fibonacci series

In typography and graphic design, layout grids are seen as aesthetic structuring systems and are based for the most part on a horizontal and vertical coordinate system. They are used to systematically structure text, images, colour, shapes, and two-dimensional space in a manner that is consistent with the purpose of the design. The Fibonacci numbers can be used to create to layout grids of various sizes and proportions.

Canons of page construction based on the Fibonacci series

The layout grids used in book design are called ›canons of page construction‹. This term refers to a page's printed areas (for example, text columns and pictures) and unprinted areas (including the top, bottom, inner, and outer margins). Canons also define line spacing and width (to ensure that pages are ›register-true‹), as well as the placement of column headings and page numbers.

The design of basic page construction canons is an ancient practice dating back to the first century, when it was part of the standard skill set of any calligrapher or scribe. Ever since the incunabula period (1450–1500), canons have been constructed from grids, which can use proportions based on the Fibonacci numbers. This produces proportional, square-unit grids, which are used as graphic layout systems for aesthetically structuring the printed and unprinted areas of book pages (typically arranged in a double-page spread).

The myth of the golden ratio in type design

The field of type design, today a branch of typography, abounds with countless theories and claims that spread and exploit the myth of the golden ratio.

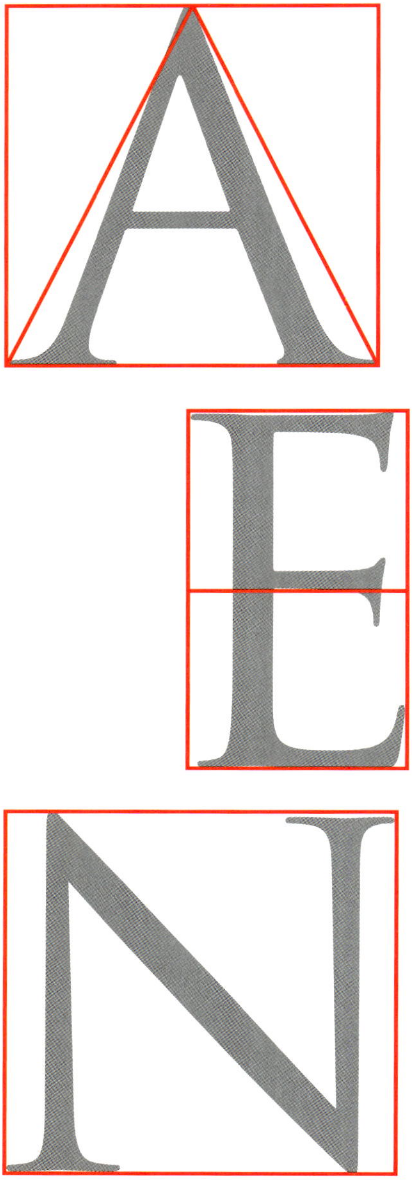

Viewed in two lines, Roman capitals consist of straight lines and curves and of basic squares, triangles, and circles. The example features Trajan Pro from Linotype.

One myth that continually resurfaces in typographic literature and teaching is that the Trajan font (113 CE), the most famous and beautiful example of *capitalis monumentalis*, was designed in accordance with the golden ratio.

In truth, however, typeface experts currently believe that both the foot of the Trajan stem, the stem itself, and the capital letters were designed in accordance with Archimedes' proportions for the cone, sphere, and cylinder. These are, respectively, 1:2:3 (cylinder of radius r and height 2r, sphere of radius r, and a straight cone of radius r and height 2r). Transferred to two dimensions, this is not the golden ratio, but instead the square, circle, and triangle – the basic shapes of our Latin alphabet.

Contemporary type designers, such as the German graphic designer and typographer Kurt Weidemann (1922–2011) or Swiss type designer Adrian Frutiger (1928–2015), also liked to reference the golden ratio, or the Fibonacci series, in their designs.

Kurt Weidemann, for example, used the golden ratio as a fanciful promotional argument for the proportions of his font family ›Corporate ASE‹, which he developed as a corporate typeface for Daimler Benz in Stuttgart. When the letters of his Antiqua typeface trilogy are subjected to typometric analysis, however, it is very difficult to find a connection between the letter architecture and the golden ratio.

Strictly speaking – from a mathematical perspective, that is – not all the claims made by type designers are true. As non-mathematicians, though, we can adopt a more relaxed attitude and humour these ideas as more or less just pleasant fantasies.

In reality, however, our Latin letters are primarily comprised of straight and curved lines, together with basic squares, triangles, and circles – and that is as true today as it always has been.

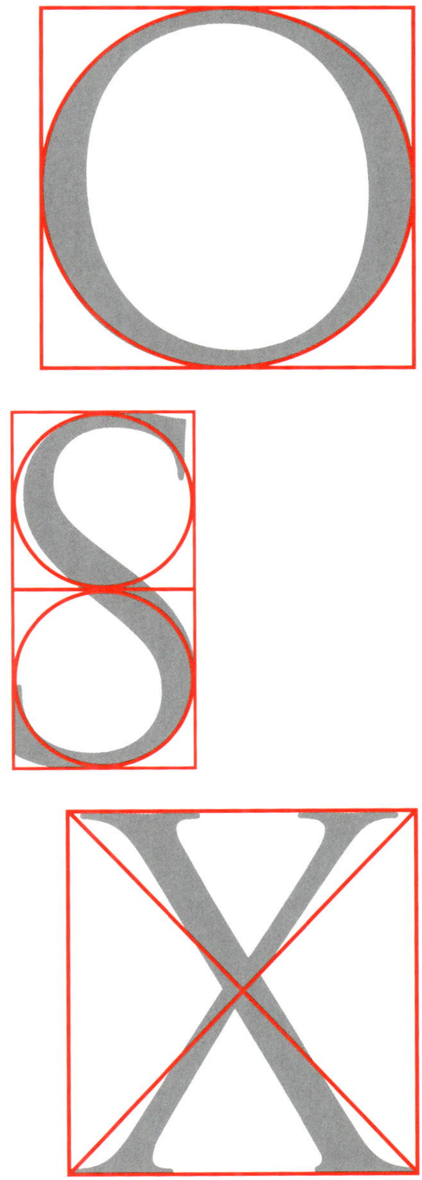

Boxes, pallets, containers
Industrial standards

Peter Hirschmiller

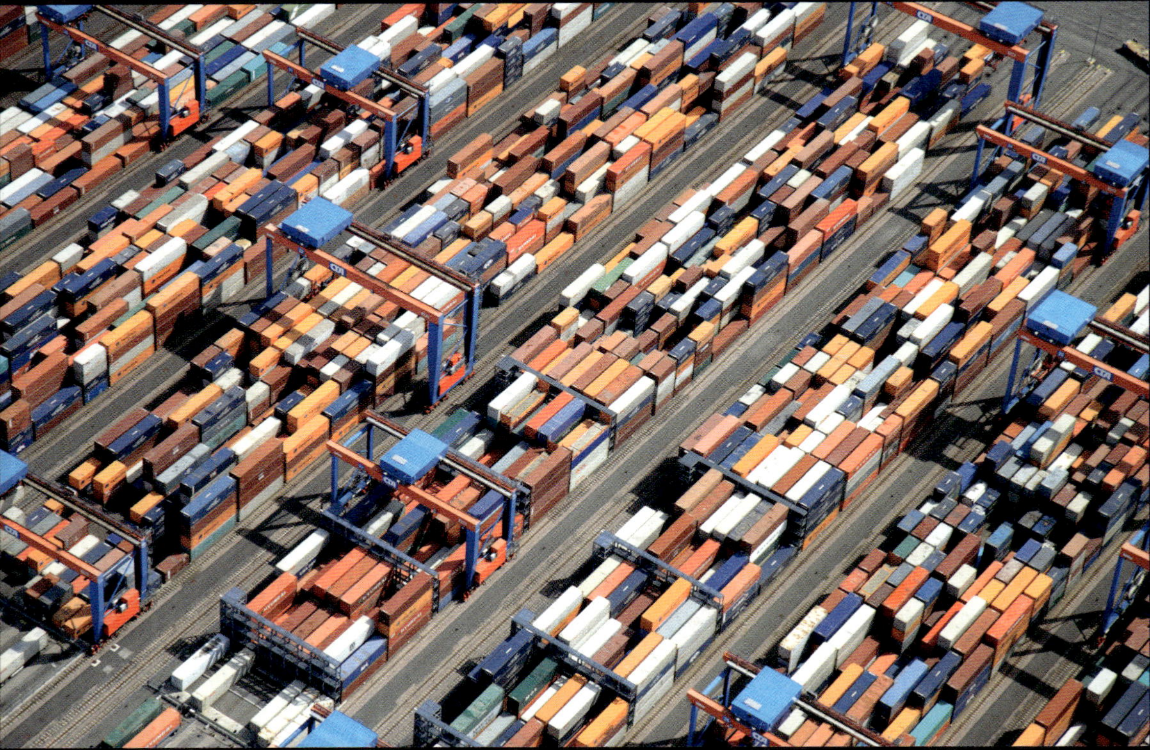

Bird's-eye view of containers at the harbour in Hamburg

Just as in art, design or architecture, it is hard to overestimate the crucial role played by proportions in international trade – though, admittedly, the requirements here are rather different. The dimensions of objects in industry are less directed to beauty and aesthetic value than, first and foremost, to standardisation and rationalisation. And this is something we encounter time and again in our daily lives – from the milk carton in the fridge to the banana box in a supermarket and the screws in the tool box, the list of standardised products seems endless. Yet it is not only the packaging or products which are subject to specific and internationally agreed guidelines. Today, the present volume of goods traffic around the world would be impossible to maintain without universally valid standards.

In 1961, the International Union of Railways (*Union Internationale des Chemins de Fer* – UIC) drove forward the development of standardised load carriers for European freight transport. The UIC set out to design a pallet optimising storage space in goods trains and, at the same time, simplifying the complex process of transferring goods from road to rail. Under European Standard EN 13698-1 and the provisions in UIC-Code 435-2 and 3, all euro pallets, or more precisely pallets within the European pallet pool, must be 120 cm long, 80 cm wide and 14.4 cm high. These pallets are comprised of nine wooden blocks and eleven boards, with the individual wooden elements fixed by 78 special nails in a prescribed nailing pattern. Depending on the moisture content of the wood, each pallet weighs around 20 kilos and has a load capacity of at least 1000 kilos. Thanks to the position of the bottom bearers on either side, these are four-way pallets, optimised for fork lift trucks.

In the meantime, the euro pallet's dimensions have influenced the architecture of entire warehouses – just as they have goods wagons or articulated trucks. This standardisation has allowed the loading times of goods wagons and trucks to be reduced by around 90 % – a massive saving in time.

The umbrella European Pallet Association (EPAL), set up in 1994, not only assures the quality of EPAL pallets, but also grants licences for euro pallet production and monitors adherence to UIC standards. With estimates of 500 million euro pallets in circulation, logistics experts talk with good reason of the euro pallet as a quantum leap in freight traffic.

However, the pallet's apparent universal usefulness ends at Europe's borders. When this ›logistic marvel‹ reaches the harbours, it meets the undisputed king of global goods traffic – the shipping container. But since this container is based on the measurement system of feet and inches common in the Anglo-American world, filling it with metric pallets optimised for Europe inevitably leaves storage space unused. As a result, it is necessary to transfer the goods at these freight interfaces. The exception here is the so-called ›land container‹, which allows euro pallets to be seamlessly stored.

The shipping containers can also produce some impressive statistics. In 2015 alone, Hamburg harbour handled 90.6 million tonnes of goods in 8.8 million containers. The idea of standardised intermodal steel containers to transport goods is attributed to American trucking entrepreneur Malcom McLean in the 1930s, although the sceptical industrial lobby initially refused to bite. Yet by the time the first container ship set out on its maiden voyage in 1956, the advantages of the container system were only too obvious. Previously it often took a week to unload and load a freighter – a task McLean's workers could accomplish in just a few hours, with a comparable saving in costs. This heralded an unbeatable rationalisation of freight transport, and McLean's former critics had to unreservedly recognise the success of the ›container revolution‹.

In 1964, the International Organisation for Standardisation (ISO) established the standard dimensions of the ›ISO container‹. Under ISO 668, these containers measure 8 feet (c. 2.44 m) in width, 8 feet 6 inches (c. 2.59 m) in height and 20 feet in length (c. 6.06 m), leading them to become known as twenty-foot equivalent units, or TEUs for short. The largest container ship sailing the world's seas at present can transport nearly 19,200 TEUs – and freighters able to carry 20,000 TEUs are already under construction.

Ultimately, whether the euro pallet with its aspect ratio of 1:1.5 is also inherently ›beautiful‹ lies in the eye of the beholder. In any case, the pallets now withdrawn from service are much sought after, and given a new lease of life as chairs, cupboards, tables, beds or even art works. But containers are also co-opted for other purposes entirely – re-interpreted as minimalist homes or mobile exhibition spaces.

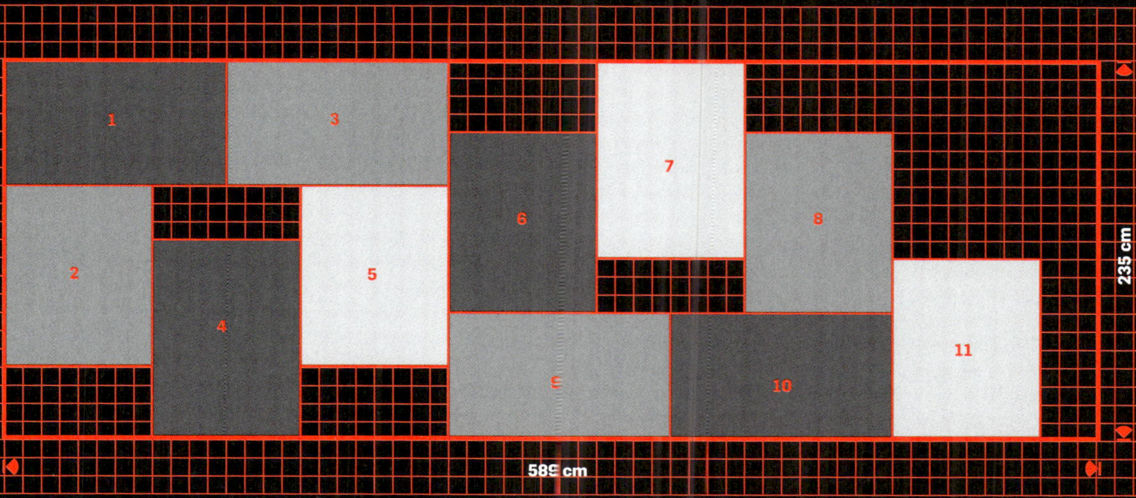

Divine details
The portrait of Luca Pacioli

Mara-Lisa Kinne

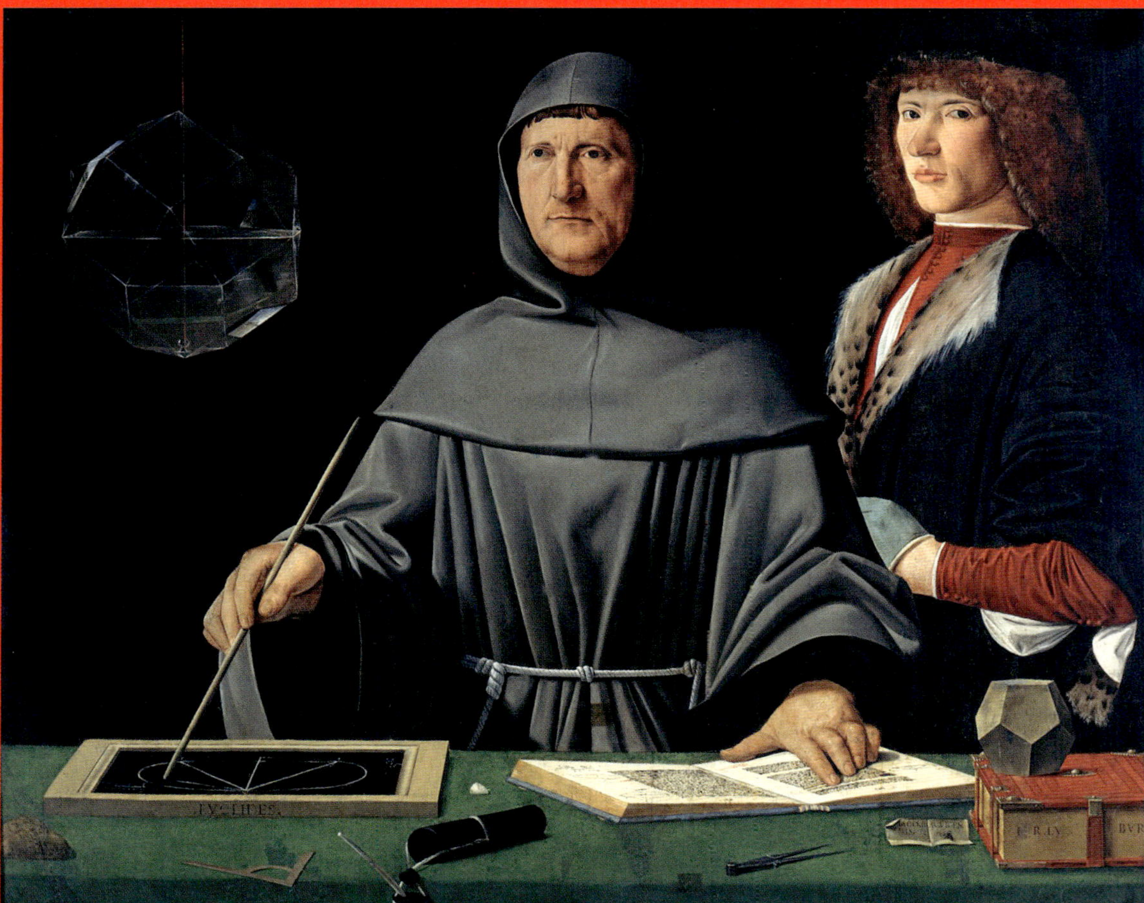

Where would one expect to find the golden ratio in art if not here? Not only does the work depict a wealth of mathematical objects, but first and foremost the centre is dominated by a portrait of the Franciscan monk, mathematician and scientist Luca Pacioli (c. 1445–1517). In 1509, Pacioli published his *De divina proportione*, an entire treatise devoted to the golden ratio.

This work, painted 14 years before Pacioli published his treatise, is most likely by Jacopo de' Barbari. It shows Luca Pacioli, then around 50 years old, together with the young Duke of Urbino, Guidobaldo da Montefeltro. The two men are standing in a darkened room behind a desk equipped with numerous objects which Pacioli required for his mathematical work. Pacioli has Euclid's major work the *Elements* open in front of him. His left hand is still resting on the book, marking a particular passage, while he points with the rod in his right hand to a geometric drawing on a slate illustrating the problem described in the text. On the table, Pacioli also has other implements required for his work – a piece of chalk, a sponge, a protractor, a compass and an ink pot. The magnificently bound red book on the right is Pacioli's main work *Summa de arithmetica, geometria, proportioni et proportionalita*, published in 1494, the year before this portrait was painted. The *Summa de arithmetica* was a compendium of the mathematical knowledge of Pacioli's day and, due to its wide circulation, it was to create the basis for his fame. The wooden dodecahedron on the book has its counterpart in the spectacular glass solid with twenty-six semi-regular faces suspended from the ceiling. The latter, known as a rhombicuboctahedron, is only held by a delicate thread fixed to the base, even though it contains several litres of water which must have been quite a weight. While Pacioli, set in the centre of the composition, is pondering a mathematical problem surrounded by the instruments of his work, the accompanying figure, set at an angle behind him, is gazing out observantly at the beholder.

Portrait of Luca Pacioli, attributed to Jacopo de' Barbari, 1495

Luca Pacioli as an admirer of the golden ratio

This is a remarkable painting for several reasons – not least due to its status as the first portrait of a mathematician believed to have been painted during the person's lifetime. Yet the inconsistencies in the work raise any number of questions. For example, what is the nature of the unfinished line on the slate which Pacioli is pointing at rather than drawing, and which is not described in the passage in Euclid? Or how should we read the figure hardly visible below the water surface in the glass solid? But other questions would also include, for instance, how far religious levels of meaning play a role, or how mathematics and painting relate to one another in this work – and here, the question of the role of the golden ratio seems especially pertinent.

Pacioli's treatise *De divina proportione* was the most important publication on the golden ratio since Euclid's *Elements*. In an approach similar to his *Summa de arithmetica*, Pacioli primarily compiled existing knowledge. Nonetheless, rather than solely addressing mathematicians, Pacioli wrote in the local *volgare*, and disseminated his work as a printed book. He emphasised the far-reaching relevance of the golden ratio, noting among other things that »without knowledge [...] of the said ratio, it is impossible to reach an understanding of much of the wonder of the most sublime things, either in philosophy or in any other science«.[1]

While Euclid only describes the golden ratio as a ›continuous division‹, Pacioli refers to the golden ratio as the ›divine proportion‹, and offers five reasons to support his claim in the fifth chapter of his book. First of all, he argues, just as God is absolutely unique, so too is the golden ratio; secondly, similarly to God as a unity of the Holy Trinity of Father, Son and Holy Ghost, the golden ratio is also a unity of only three elements; thirdly, God must remain ultimately indefinable, and the golden ratio is also impossible to define by any rational quantity; fourthly, irrespective of the value used for this proportional ratio, it remains invariable, just as God is in essence immutable; and finally, Pacioli adds as the fifth reason, just as God »conferred being to celestial virtue« as the fifth substance alongside the four elements, the golden ratio is necessary to define the dodecahedron as the fifth figure in the set of regular bodies. With the golden ratio so theological charged, it is fitting for Pacioli to pursue his mathematical studies here dressed in his Franciscan robe with his head covered by his hood which, according to Hannah Baader, can be read as a sign of contemplation.

Given that Pacioli propagated the idea of the golden ratio and wrote his treatise in 1498, although it was first published in 1509, the obvious next question is whether the golden ratio also plays a role in this painting. And if so, in what form?

Scholars on the trail of the golden ratio

In search of the golden ratio here, one can find numerous clues and pointers. Simply due to the dodecahedron's mathematical qualities, it could be seen as alluding to the golden ratio. And although Euclid's Proposition XIII, 8, which Pacioli has also apparently just read in the *Elements*, is not directly related to the golden ratio, the following Proposition XIII, 9, which can be seen on the next page of the book, places the length of the sides of a hexagon and decagon inscribed in the same circle in relation to the golden ratio. For the drawing on the slate, Enrico Gamba has offered a number of very complex possible ways of completing the sketch so it would relate to the golden ratio. In this context, Heinrich Emil Timerding's suggestion that the horizontal water level line in the rhombicuboctahedron divides the two edges closest to the beholder in the golden ratio appears to be more compelling. An observation on the simple sum on the slate is even more striking. There, the numbers 478, 935 and 621 are correctly added to make 2034. In their recent paper, though, Renzo Baldasso and John Logan have suggested a new way of reading these numbers. Following Leonardo's practice of mirror writing, they read these digits in the reverse sequence, producing the numbers 874, 539 and 126. If these numbers are then set in relation to each other as pairs, they produce powers of ϕ: 874:539=1.62152..., which is equal to the value of ϕ, 1.61803..., with an error variance of less than 1 %. The second pair of 539:126=4.27778..., closely approximates to the third power of ϕ, 4.23604.... Moreover, the quotient of 874 and 126, which is 6.9351..., is similar to 6.85403, the value of the fourth power of ϕ. Could this just be a coincidence?

1 Luca Pacioli, Divina Proportione. Die Lehre vom Goldenen Schnitt, Constantin Winterberg (ed.), Vienna, Graeser 1889, p. 171.

The divine proportion as a key issue

In this case, two things are remarkable. First of all, the various possible references to the golden ratio do not immediately suggest themselves to the viewer, but have to be discovered and deduced. Second, most of the ideas mentioned have little to do with the painting's visual qualities. Where such considerations are triggered by the qualities of an object, a text, geometrical drawing, or calculation, only the fact of their depiction is crucial and not how they are depicted. Furthermore, the format of the painting itself is not a golden rectangle, nor does dividing the sides in the golden ratio produce any significant meaning. Although one of the four lines does touch the lowest point of the rhombicuboctahedron, they do not otherwise mark any particular points in the composition.

This inconsistency matches both the painting in its entirety as well as Pacioli's treatise. With its various allusions to mathematics, the painting is clearly supposed to be a work in a ›mathematical style‹, yet it does not even have a unified vanishing point for its construction of perspective. In a similar way, even though Pacioli's title of *De divina proportione* seems to imply that proportions are useful in philosophy, perspective, painting, sculpture, architecture and other mathematical subjects, in those sections dedicated to the proportions of the human body, architectural elements or letters, he hardly mentions the golden ratio at all.

So here, in a similar spirit, the golden ratio is not the subject of the painting, but only relevant at certain points. Nonetheless, this portrait does encourage the attentive viewer to engage with the mathematical content – as is apparent, not least, in the various attempts to prove the presence of the golden ratio in this work.

Baader, Hannah: »Das fünfte Element oder Malerei als achte Kunst. Das Porträt des Mathematikers Fra Luca Pacioli«, in: Valeska von Rosen/Klaus Krüger/Rudolf Preimesberger (eds), Der stumme Diskurs der Bilder. Reflexionsformen der Ästhetischen in der Kunst der Frühen Neuzeit, Munich/Berlin, Deutscher Kunstverlag 2003, pp. 177–203.

Baldasso, Renzo/Logan, John: »Between the Golden Ratio and a Semi-Perfect Solid: Fra Luca Pacioli and the Portrayal of Mathematical Humanism«, in: Ingrid Alexander-Skipnes (ed.), Visual Culture and Mathematics in the Early Modern Period, London, Routledge 2016 [in press].

Baldasso, Renzo: »Portrait of Luca Pacioli and Disciple: A New, Mathematical Look«, in: The Art Bulletin, 92 (1–2), 2010, New York, College Art Association, pp. 83–102.

Ciocci, Argante: Luca Pacioli e la matematizzazione del sapere nel Rinascimento, Bari, Cacucci 2003.

Fredel, Jürgen: Maßästhetik. Studien zu Proportionsfragen und zum Goldenen Schnitt, Hamburg, Lit 1998.

Gamba, Enrico: »Proviamo a rileggere il ›Doppio ritratto‹ di Luca Pacioli«, in: Francesca Maria Cesaroni et al. (eds), Le tre facce del poliedrico Luca Pacioli, Urbino, Centro Internazionale di Studi Urbino e la Prospettiva 2010, pp. 81–97.

Kinne, Mara-Lisa/Krischel, Roland/Neuner, Stefan: »Annäherung an ein Rätsel. Das Bildnis des Luca Pacioli von Jacopo de' Barbari« in: Wallraf-Richartz-Jahrbuch 76 (2015), Cologne, Dumont, pp. 79–115.

Pacioli, Luca: Divina Proportione. Die Lehre vom Goldenen Schnitt, Constantin Winterberg (ed.), Vienna, Graeser 1889.

van der Schoot, Albert: Die Geschichte des Goldenen Schnitts. Aufstieg und Fall der göttlichen Proportion, Stuttgart-Bad Cannstatt, frommann-holzboog 2005.

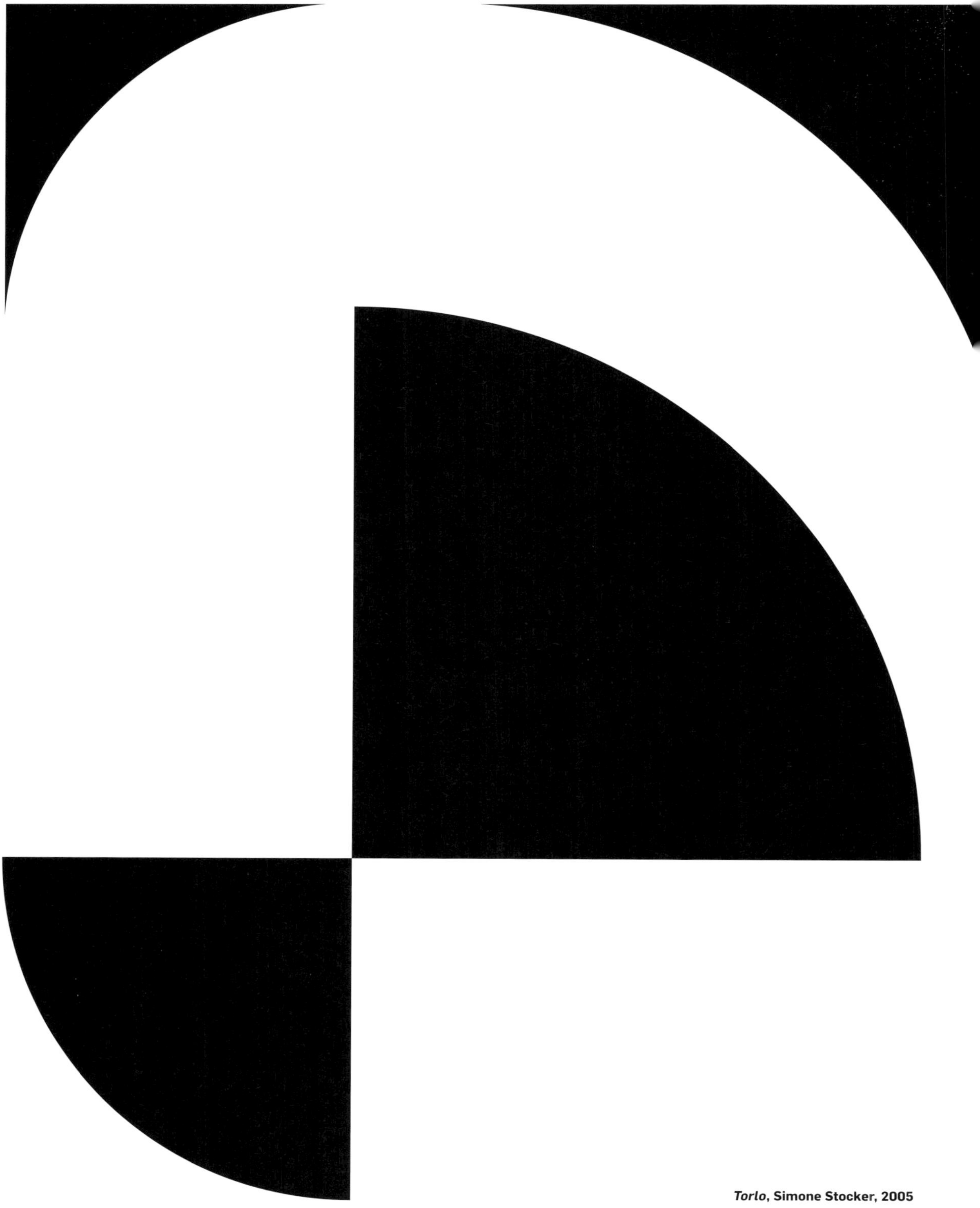

Torlo, Simone Stocker, 2005

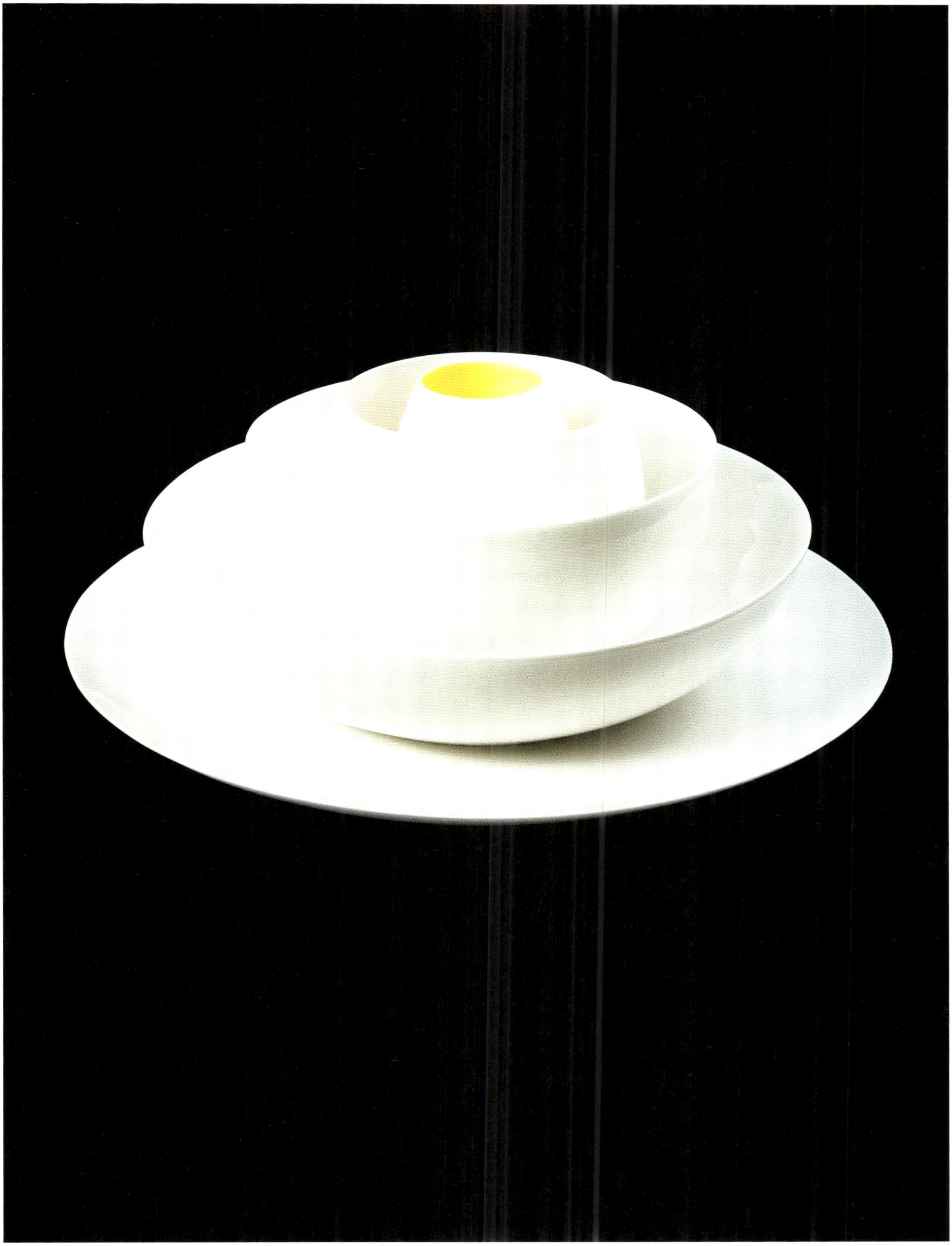

20 steps around the globe …
A Land Art project by Jo and Tuula Niemeyer

Timo Niemeyer, Daniela Tisevski

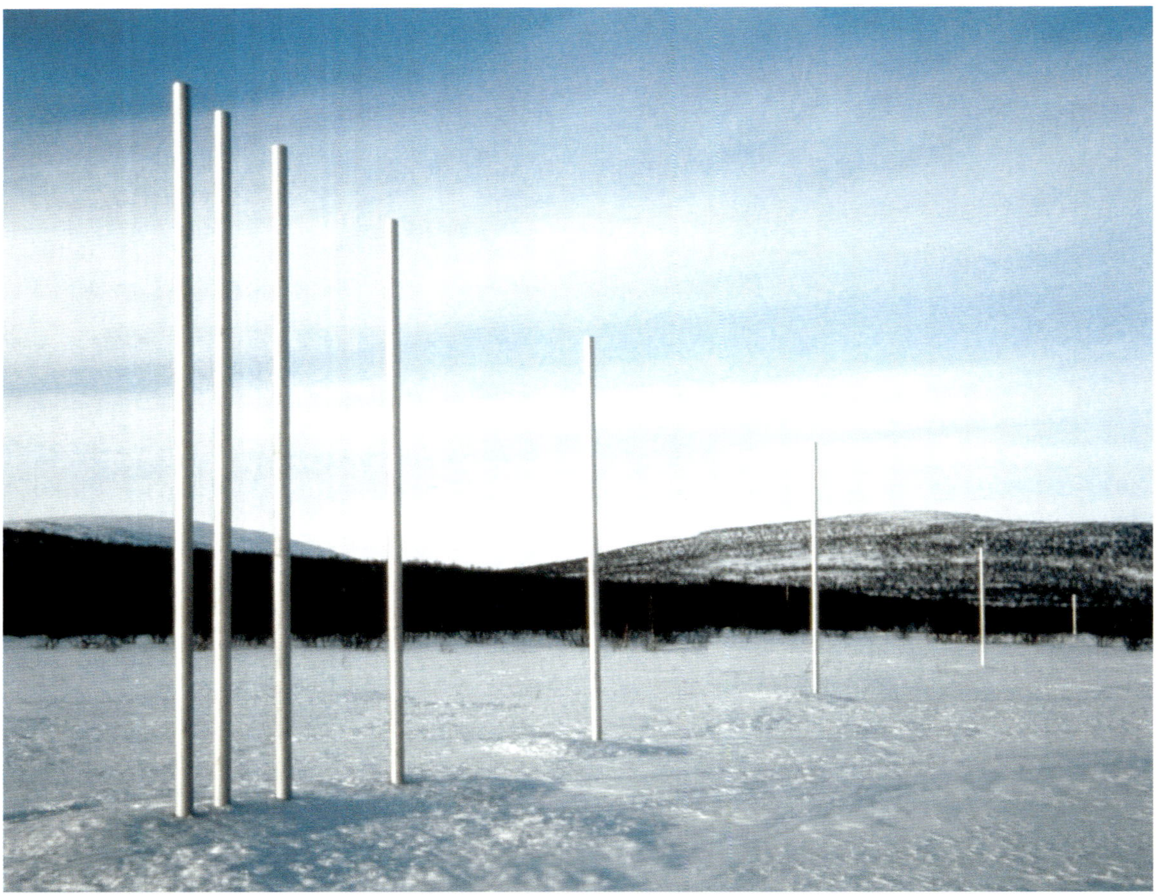

Although the concept of a spherical Earth was already prevalent in ancient Greece, the planet's dimensionality continues to fascinate people even today. The abstract idea of living on a sphere has to be mastered in many everyday situations – whether admiring a sunset (or, more rarely, a lunar eclipse) or gazing dreamily at the distant horizon of the sea. Pythagoras, Plato and Aristotle were convinced that the Earth was spherical, but it was not until the second century BCE that the Greek scholar Eratosthenes, head of the legendary Library of Alexandria, succeeded in calculating the Earth's circumference. His results were astonishingly precise, accurate to the nearest 300 km.[1] Visiting the oasis of Syene on the Nile (today's Aswan), Eratosthenes had noticed the reflection of the midday sun in a deep well. As he looked, he was struck by the brilliant idea of using the sun's angle of incidence to determine the circumference of the Earth. Even though the mathematical computation of the globe's circumference was achieved in antiquity, today we still find it difficult to comprehend the dimensional finitude of our planet. Again and again, astronauts, cosmonauts, taikonauts – all those who have had the privilege of observing the Earth from orbit – recount their amazement at how comparatively small and fragile our planet looks from space. Their wonder is not without an element of criticism of human beings' huge impact on nature and the climate through consumption and wars that jeopardise our unique ecosystem.[2] If the totality of the planet were easier for human beings to grasp intellectually, they might be able to treat its limited resources with more care.

Figure 1. Steps 1 to 8 in Ropinsalmi

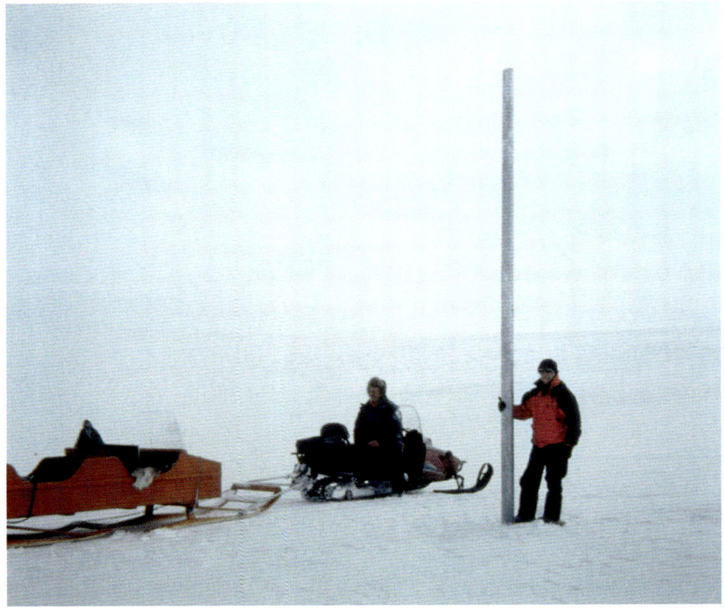

Figure 2. Installing the pillars at -30°C

Nature and humankind united in art

It was to make the globe experienceable in this way that Jo Niemeyer created his world-encompassing Land Art project *20 Steps*. Born in 1946 in Alf on the River Moselle, Niemeyer is associated with Concrete Art, a movement with close links to mathematics and the natural sciences. His works leave nothing to chance: every line, surface and colour scheme is subject to clear mathematical rules. Niemeyer constructs his works on the basis of the golden ratio, and is completely committed to the exploration of proportions and structures. Colour thus plays only a subsidiary role; its effects are there solely to emphasise the forms. Accordingly, in Niemeyer's works we find only the pure primary colours red, blue and yellow, along with the non-colours black and white. In his more recent works, Niemeyer has narrowed his colour palette even further, using only red, black and white. The cornerstone of Jo Niemeyer's work as an artist is his determination to generate »a dialogue between humanity and nature« – a connection between art and the environment, between artistic action and experimental endeavour.[3] In this sense, Land Art, as the transformation of geographical space into a work of art, is a logical extension of Niemeyer's work in the traditional media of paper and canvas and his sculptures.

Land Art emerged in the United States in the 1960s, differing from Minimalism primarily in its component of explicit social critique – Minimalism, striving for objectivity, was mainly to be found in the context of galleries and museums. Many proponents of Land Art regarded contemporary visual art as driven by commercial speculation. Refusing to supply the capitalist apparatus with new consumer goods, they wished to create art that would unfold only within the landscape, free and not for sale.[4] Rather than inserting the works into the environment like sculptures, thus relegating the environment to the role of a backdrop, Land Art aimed to integrate art as such into the landscape.

20 steps around the globe

This principle of Land Art also underlies Niemeyer's concept for *20 Steps*, which systematically divides the globe into twenty individual, incremental segments or steps (Fig. 3, pp. 116/117).

The formula used for their progression is the golden ratio, something that permeates the whole of Niemeyer's oeuvre. In its mathematical expression, the golden ratio is considered to epitomise proportionate harmony and aesthetics in the realms of art, design and nature.[5]

In *20 Steps*, the golden ratio is used to divide up the full circle (in other words, the globe). This yields the golden angle, 137.5...°, found in botany in the spiral arrangement of leaves as described by phyllotaxis.[6] To identify the exact location of the future sites in different countries and continents, Niemeyer used GPS navigation and a computer program specially designed for the project.

Figure 3. Schematic sketch for 20 Steps

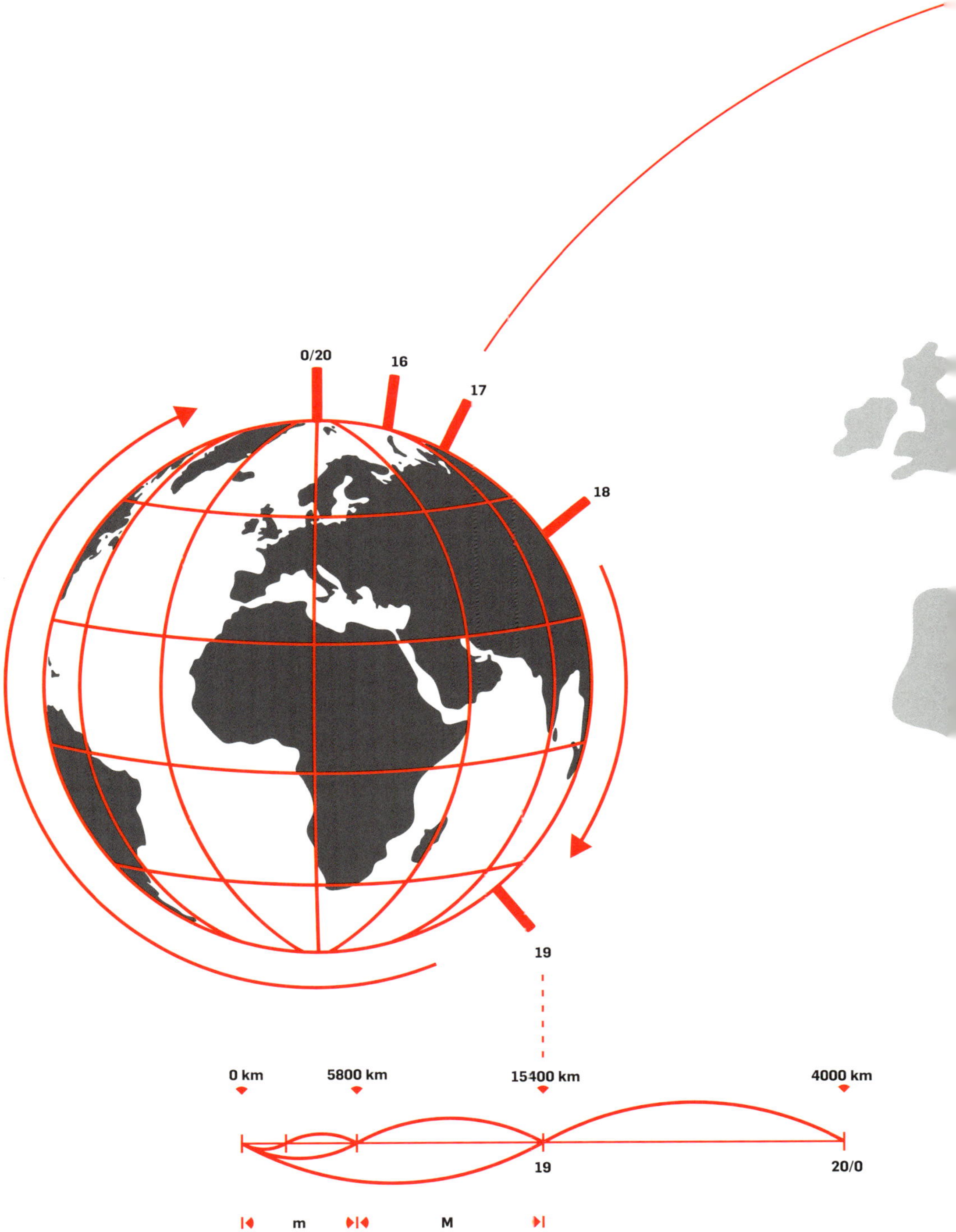

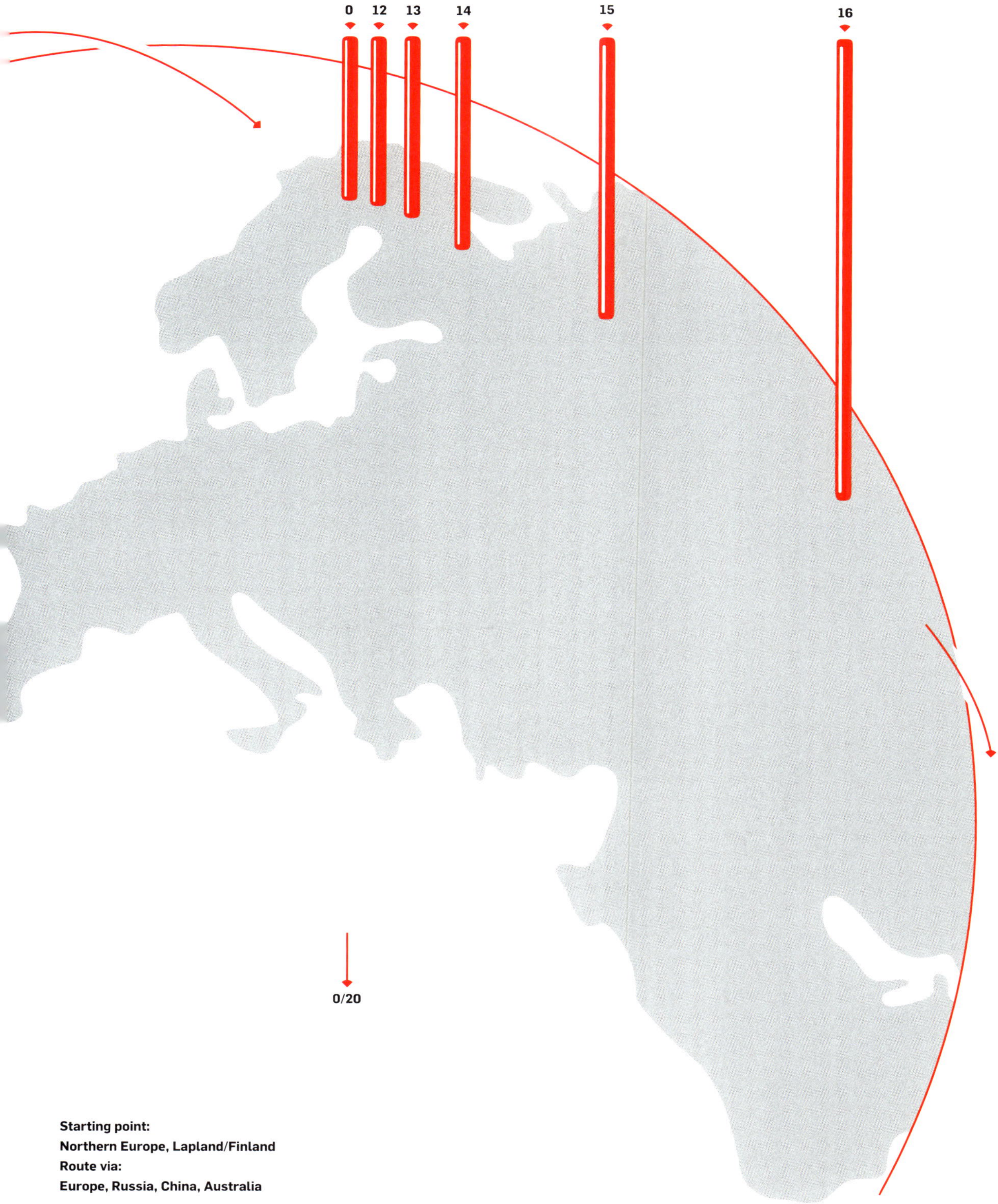

Starting point:
Northern Europe, Lapland/Finland
Route via:
Europe, Russia, China, Australia

The first step is only 45.8... cm long. Calculated on a precisely defined ratio between two segments whereby the shorter segment (›minor‹) is to the longer segment (›major‹) as the major is to the whole (minor plus major), the length of the second step is 119.8.. cm. Looking at the first two segments and noting the proportion in which they stand to one another, it is almost impossible to imagine that a mere eighteen further steps on the same pattern will be needed to circumnavigate the whole globe. But from its start point, the series develops in a progression that is dynamic and logarithmic – defined mathematically on the basis of the golden ratio – and so rapid that only the first few pillars remain within sight (the third pillar is already 3.13... m away).[7] The first ten steps, maybe eleven, are still visible for the viewer, and they mark the installation's greatest density. The point is not to see more or see less, however – it is to perceive the world in a new and different way.

From macrocosmos to microcosmos

The progression can begin from any point on the Earth, because there is no ›point zero‹, either philosophically or mathematically. ›Our world‹ is somewhere in the middle of things, in a logarithmic spiral whose interior leads us into the microcosm, its exterior into the macrocosm. The first pillar does not only divide the Earth or the organism's cell at this point – even the smallest atom in its ground state can be divided further, in a voyage into the microcosmos.[8] We reach the microcosmos when the first step in Niemeyer's installation, the 45.8... cm between the start point and the second element, is divided again into another twenty steps using the golden ratio. In a similar way to Niemeyer's project itself, the first eight of these diminishing steps can be perceived by the naked eye. They correspond to the measurements of the human arm, hand, and so on down to the size of a tiny grain of sand. From then on, we enter the world of cells and molecules, and a microscope is needed to view the work. We continue to move towards the point of departure, encountering atomic orders of magnitude.

The distance between the last step and the start point is more or less exactly the size of a hydrogen atom, ten millionths of a millimetre.

The variations of the journey from point 1 into the infinity of the microcosmos or macrocosmos recall the words of Anaxagoras in the fourth century BCE: »Nor is there a least of what is small, but there is always a smaller; for it cannot be that what is should cease to be by being cut. But there is also always something greater than what is great.«[9] As Anaxagoras' comment shows, such journeys are certainly nothing new in the imagination of scientists and artists. Philip and Phylis Morrison exemplified this with a visual voyage through the micro- and macrocosmos in *Powers of Ten: A Book about the Relative Size of Things in the Universe*.[10]

Niemeyer's Land Art project was realised in 1997. It started from Ropinsalmi, a village in Finnish Lapland near the Swedish border and north of the Arctic Circle. After approaching the local Finnish, Norwegian and Russian authorities and – most importantly – the Sámi, the indigenous inhabitants of this harsh tundra, Jo and Tuula Niemeyer obtained official permission for the temporary installation of pillars, each 5.5 m high, to mark the twenty steps. Jo Niemeyer had planned the route to run in a north-easterly direction at an angle of 74.54°, partly in order to reach as many of the steps as possible by land. The first ten points were visible to the unaided eye, as was the eleventh point if visibility conditions were good. Point 12 was still on Finnish territory. Points 13 and 14 were in the district of Kautokeino, Norway; point 15 in Russia near Nikel and the mountain Gora Shuort.

In other words, the initial points were all located on rough and hard-to-reach terrain. Today, it is becoming increasingly easy and convenient to cover even the longest distances on our Earth, which was reason enough for Jo and Tuula Niemeyer to travel the first fifteen steps (all on European territory north of the Arctic Circle) on foot, ski or snowshoe, using sledges to transport the materials (Fig. 2, p. 114). When the first pillars were erected, temperatures often were below -30°C – but the inimitable setting in an austere, snowy tundra repaid the arduous work of installation. Point 16 reached the Barents Sea, and point 17 the area south-west of Novy Urengoy in Siberia. Point 18, in the Inner Mongolian city of Baotou (People's Republic of China), was very close to the Great Wall of China and the Yellow River. Point 19 was in the south of Australia, around 60 km south of Melbourne in the Wilsons Promontory National Park. The route continued across the Antarctic and South America, via the North Atlantic; the final step was then united with the first at the starting point in Ropinsalmi (Fig. 1, p. 112). Beginning and end became one – an emblem of the infinity of our universe, the macrocosm. The implanting of the first fifteen pillars was documented on film and in photographs, and the short film *20 Steps Around the World* was released in summer 1998.[11]

20 Steps may be regarded as an interdisciplinary expedition into the borderlands between science, art and philosophy, zones that are largely neglected in scholarship today. In an era marked on the one hand by the triumph of technology, but on the other by the burgeoning of fanaticism, war and threats to the ecosystem, our planet itself seems the object best suited to making space and time – the here and now – experienceable. The installation makes it possible to apprehend the Earth's size and intensifies our awareness of it.

1 Manfred Vogel, Archimedes und der LKW, Hamburg, Verband Dt. Schullandheime 2000, p. 176.
2 Florian Hildebrand, »Satellitenbilder – mehr als tausend Worte«, Globale Urbanisierung: Perspektive aus dem All, in: Hannes Taubenböck et al. (eds), Heidelberg, Springer Spektrum 2015, p. 228.
3 Felicity Shaw, »Von Schönheit und Proportion«, in: Mathematikum Gießen (ed.), Jo Niemeyer – Moderne Mathematische Kunst, cat. Mathematikum Gießen, Gießen 2013, p. 16.
4 Ben Tufnell, Land Art, London, Tate Modern 2006, p. 51.
5 Jo Niemeyer, »20 Steps«, in: Mathematikum Gießen (ed.), Jo Niemeyer – Moderne Mathematische Kunst, cat. Mathematikum Gießen, Gießen 2013, p. 19.
6 Ulrich Grevsmühl, 20 Punkte – 20 Steps. Ein Land-Art Projekt von Jo Niemeyer, Schluchsee, Edition Partanen 1993, 23 ff.
7 Ibid., p. 15.
8 Ibid., p. 27 ff.
9 Anaxagoras, Fragment 3, translated in: John Burnet: Early Greek Philosophy, London 1908, p. 299.
10 Philip Morrison, Phylis Morrison and the studio of Charles and Ray Eames, Powers of Ten: A Book about the Relative Size of Things in the Universe and the Effect of Adding Another Zero, Redding, CN, 1982.
11 20 Steps (film): https://www.youtube.com/watch?v=nPx3Ap81nIU (accessed: 16 June 2016).
12 Gerda Ridler, »Einführung«, in: Square, Museum Ritter (ed.), cat. Museum Ritter Waldenbuch, Heidelberg, Wunderhorn 2005, p. 27.
13 Eugen Gomringer, »Die Konkrete Kunst wächst über sich hinaus«, in: Grevsmühl, Ulrich, 20 Punkte – 20 Steps, p. 12 ff.

In *20 Steps*, Jo Niemeyer worked on a scale to which the pioneers of non-figurative art in the early twentieth century had aspired: they strove to achieve universality by drawing on non-artistic elements such as science, philosophy, industry, architecture and technology.[12] As Eugen Gomringer wrote of the installation *20 Steps*: »From its beginnings, Concrete Art has seen itself as universal in a dual sense: as being entire and as ›embracing the whole world‹. But there has probably never been a project that demonstrated Concrete Art's ability to create world-embracing, visible actuality as convincingly as Jo Niemeyer's *20 Steps* does.« Niemeyer's project, Gomringer continues, emanates the elegance of simplicity in a way that can be found only in the playful space of Concrete Art. »Is it not a task worthy of this art,« he asks, »to remind us of the proportionality of the world and on Earth?«[13]

Nothing is forever

Niemeyer's twenty steps were left standing for just a few days, enough time to document the project – for Land Art is as ephemeral as nature itself. Richard Long's circles of wood or stone are exposed to wind, weather and vegetation, as are Andy Goldsworthy's installations. Christo and Jeanne-Claude's wrapping projects come with a strict time limit, which is exactly what allows them to unfold their impermanence and their special charisma. Just like *20 Steps*, these works find parallels with human existence, and this makes them more human, as Ulrich Grevsmühl points out in his publication on the project. In this way, ephemerality comes to articulate singularity.

Grevsmühl, Ulrich: 20 Punkte – 20 Steps. Ein Land-Art Projekt von Jo Niemeyer, Schluchsee, Edition Partanen 1993.
Mathematikum Gießen (ed.): Jo Niemeyer – Moderne Mathematische Kunst, Gießen, Mathematikum Gießen e.V. 2013.
Tufnell, Ben: Land Art, London, Tate Modern 2006.

The golden system of rules meets intuition

Insights into the design process

Mark Braun, Franziska Land

A constituent feature of everyday culture today is the way one element reciprocally relates to another. In daily life, we constantly encounter the most varied combinations in approaches to relating two or more elements in terms of size, spacing and structure – from a living room with sofa, table and carpet to leafing through a print magazine, mixing and matching tops and bottoms to create fashionable outfits, or even in a restaurant's floor plan for its tables. Product designers refer to this interplay as composition, graphic artists define the relations in the layout, and fashion designers talk of styling.

Although designers across the spectrum consciously focus on the ratios of elements and defining their proportional relations – whether in a reflected critique or the casual routine of experience –, individual beholders rarely notice such intense engagement with the design task. After all, we only question objects and their arrangements when this interplay seems discordant.

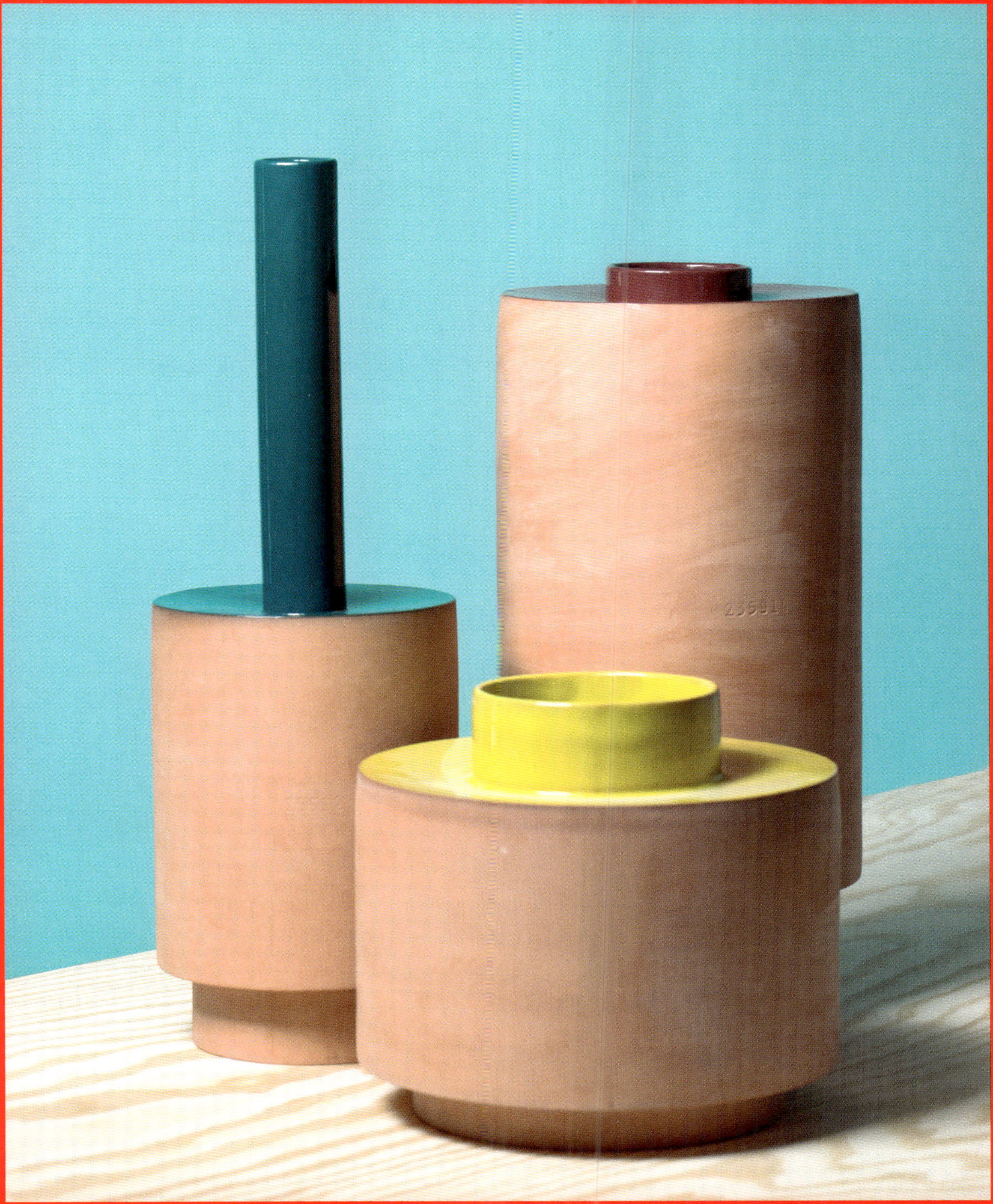

Figure 1. ***Les Trois*** **ceramic vessels, based on ancient amphorae for oil, wine and honey**

Successful aesthetics: Complex rules of proportion and associative semiotic impact

In discussing this phenomenon, it seems apposite to talk of intuition, since producing the impression of an inherently harmonious object requires a very subtle process of observation – and if a design has not achieved this satisfying final state, then ›something is still not quite right‹. Yet even so, when we look at the object, we may find it almost impossible to succinctly express those precise improvements necessary to give the object a ›finished‹ appearance. This puzzling moment of the gaze grating on the dissonance of a work is especially noticeable in the process of developing a product design, for example, if a sofa's back height does not really match the seat depth. In such cases, it is primarily our emotional response which leads us to regard particular formal relationships as incomplete and others as harmonious.

Nonetheless, this phenomenon seems to be informed by a complex and subtle set of rules. After all, we can find any number of instances of sheer aesthetic beauty in the natural world – from the sequence and structure of petals on a blossom to the snail shell's spiral shape. More precisely, these natural forms have an inherent rhythm, a recurring harmonious and rhythmical distribution

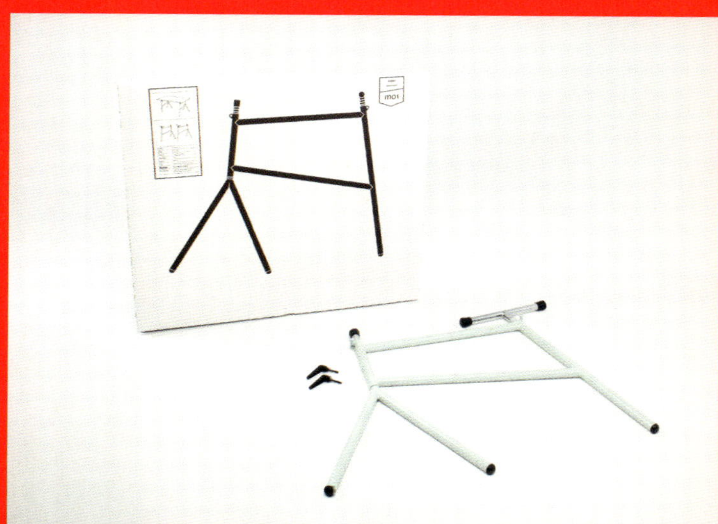

Figure 2. Trestle *m01*, a hybrid of the golden ratio and the Fibonacci sequence

of segments or surfaces that have been repeated since time immemorial. Their recurrence lends this rhythm a primal familiarity and pleasure, and this in turn has generated apparent universal agreement on aesthetic beauty and its dynamic totality – summarised as a plausible, calculable set of rules in the golden ratio or the Fibonacci sequence.

Mandatory and optional: From visible functionality to dynamic aesthetics

The aesthetics of form is a mandatory element in any successful design. In his third principle of good design, Dieter Rams enshrines this view in his statement that »Good design is aesthetic« and, in this sense, increases the user's or beholder's sense of well-being. In other words, we like to surround ourselves with beautiful products. Not only are they appealing, but they also seem familiar. Moreover, aesthetic design is more readily accessible – or more precisely, a formal design incorporating familiar proportions and volumes accelerates user acceptance. Thanks to an object's more or less familiar exterior appearance, users can grasp the immanent function faster. And indeed, where a product's form and dimensions are informed by familiar, socially defined and learned formal signs which users know from a related application context, they can deduce the product's function automatically. The visual form evokes learned patterns that users unconsciously compare with previous experiences, enabling them to functionally categorise the product. Dieter Rams has neatly summarized this semiotic effect in the words: »At best, it [i.e. the product] is self-explanatory.«

Function and aesthetics are therefore closely linked. Hence the classic product designer creating consumer goods such as furniture or small appliances needs to harmonise function and form. Here, although aesthetics runs a close second, function is the key factor since, ultimately, to quote Rams, »A product is bought to be used«.

In keeping with this prioritisation, the design process initially explores the given application context, since a product's function and specific uses essentially determine form, size and proportions. Although one might then well consider a functioning product as the end of the design process, a good designer is nevertheless keen to explore the product's formal language until a functional and aesthetic symbiosis has been achieved which, additionally, is endowed with signalling power. In this final step, proportional relationships such as the golden ratio and the Fibonacci sequence – a more complex variant with a larger range of parameters – provide reliable tools for professionally evaluating the design on a formal level and, if necessary, altering it to ensure the desired quality. The division ratios anchored in these tools offer a simple method of taking product design onto an essentially aesthetic level, to a stage of dynamic formal aesthetics whose semiotic effect also produces a sense of familiarity. At this point, though, a designer must re-critique the design to decide whether applying the golden ratio has merely rendered the product trivially harmonious and made its semiotic impact so familiar that, compared to other products, the design has lost in appeal and character.

This renewed critique is absolutely vital. The golden ratio may be a useful basis for design, yet it is far from a successful formula for such a perfect product that even casual observers are moved to pause and look. A perfectly harmonious design, with every individual detail and the overall look and feel determined by the golden ratio, may at times slip into the dull and conventional. In creating an aesthetic that is truly compelling and forges a product's identity, harmony can perfectly well cope with shades of disharmony.

Flashback: Designing the trestle m01 (Thonet) and the Les Trois series of vases (Granville Gallery)

For our design of the tubular steel trestle *m01* produced for Thonet and commissioned by the interior designer ›minimum‹, Berlin (Fig. 2, p. 124), we combined the golden ratio with Fibonacci ratios. The Fibonacci pair of 1:1 on the single rod leg is juxtaposed with the fork leg's golden ratio of 1:1.618. Optically linked by two cross struts, this encounter between these two proportional principles creates a dynamic plane which spans the trestle, and which viewers immediately associate with a bicycle frame. While the relationship of fork to rod width initially follows the Fibonacci 1:1, the fractional shortening of the ›handlebar‹ leg subsequently creates an even more dynamic overall impression. Here, then, the gaze intuitively guided by expertise exists alongside a system of rules and can fruitfully bend or break them.

In particular, less complex products such as vases call for expanding the playful approach to proportions which resonates in the *m01* design. Here, volume and size ratios, materiality and colours can have an even stronger impact on an object's appearance, as is evident in the *Les Trois* (Fig. 1, p. 123) series. These three ceramic vessels, designed for the Granville Gallery in Paris, are a modern interpretation of ancient amphorae for oil, wine and honey. The proportional relation between the vessels' bodies, bases and lips has an inherent dynamic harmony, indebted above all to the Fibonacci sequence. Yet here too, the system of rules was critically examined and adapted in detail. In this way, the basic aesthetic design which emerged can provide a cogent and harmonious model for further elements, such as the partially fired glaze with associative colour codes and the striking use of different materials. Such a powerful and distinctive formal aesthetic appearance in its totality can only be created through such an approach. In conclusion, then, good design is only possible by combining the elements of a ›golden system of rules‹, intuition and a conceptual superstructure.

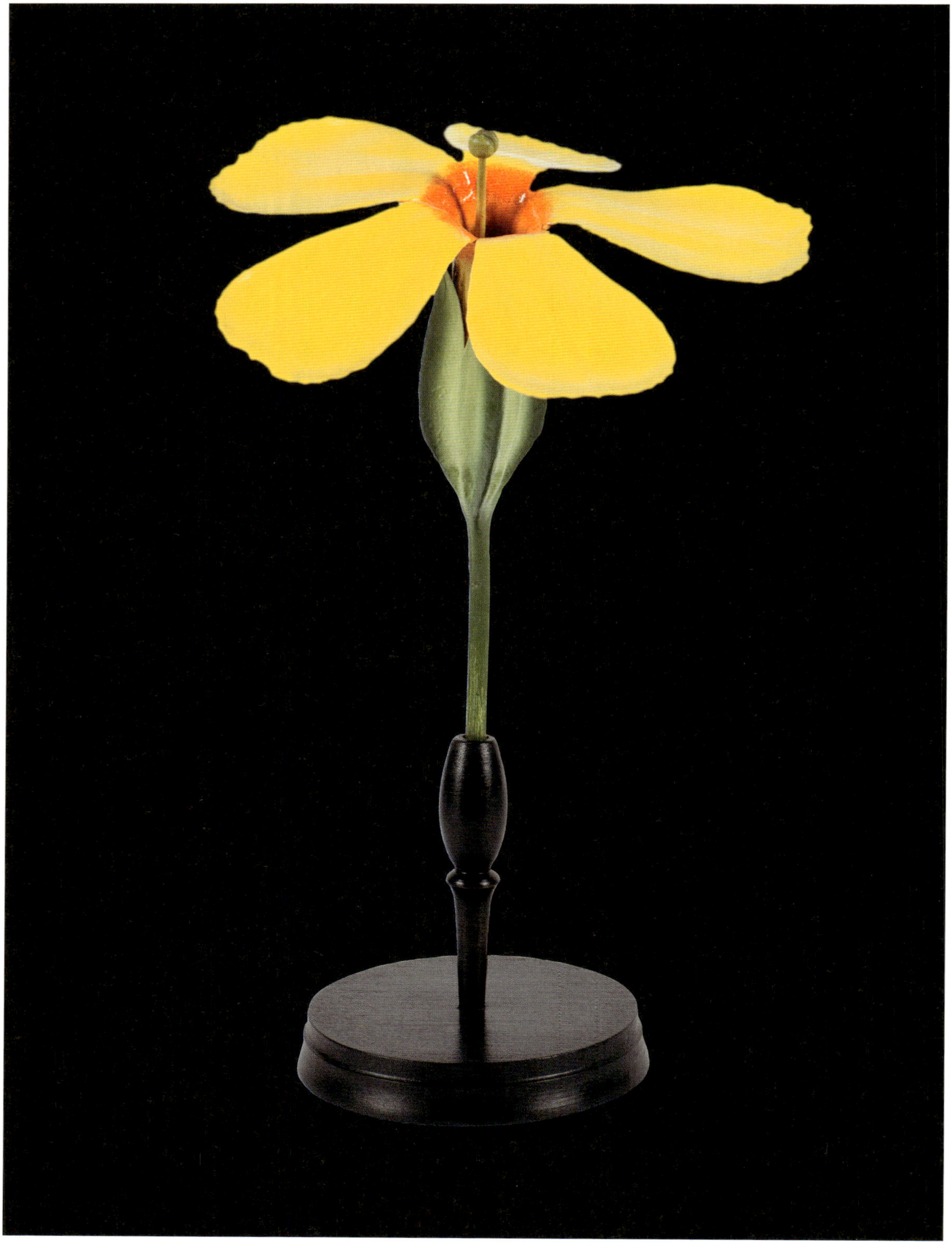

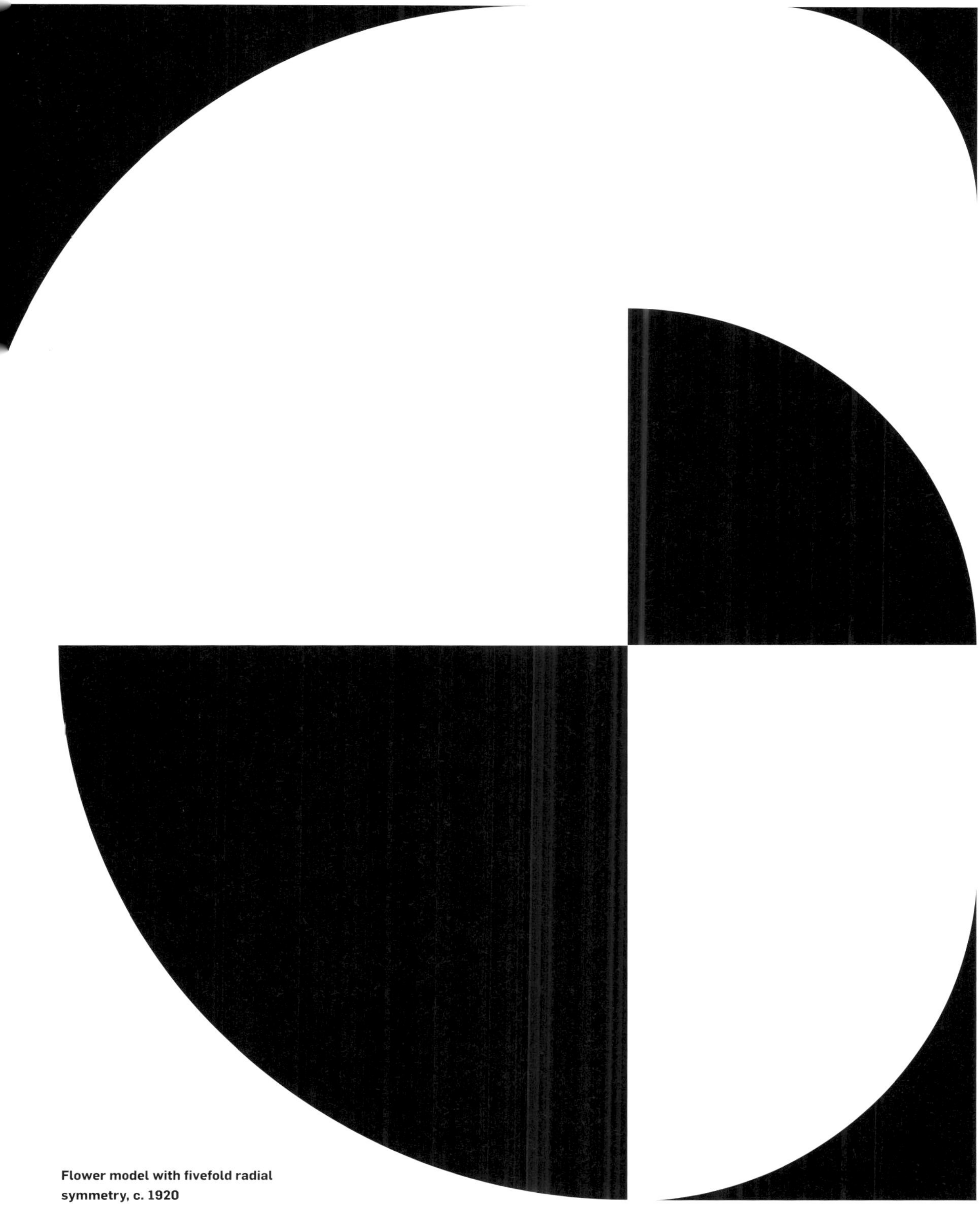

Flower model with fivefold radial symmetry, c. 1920

The attempt to measure beauty
On the history of empirical aesthetics

Katharina Fendius

Gustav Theodor Fechner and the two Madonnas

Until the first half of the nineteenth century, the Dresden *Virgin and Child with Burgomaster Jakob Meyer zum Hasen and his Family* (Fig. 1) was regarded, as described in current art historical debate, as »the major pendant to the Italian Renaissance«.[1] In 1871, physicist and philosopher Gustav Theodor Fechner even characterised this as »the main work of German painting, [and] at the same time, one of the favourite pictures of the German nation«.[2] From 1855, underlining such a claim, the panel painting was displayed facing Raphael's *Sistine Madonna* in the central exhibition space in Dresden's new *Gemäldegalerie*. In 1821, when Hans Holbein the Younger's original work, later known as the *Darmstadt Madonna* (Fig. 2, p. 132), was rediscovered, it provoked a controversy over the two works which, as the »founding moment in an academic art history,«[3] proved to be constitutive for this scholarly subject which was then in a process of differentiation.

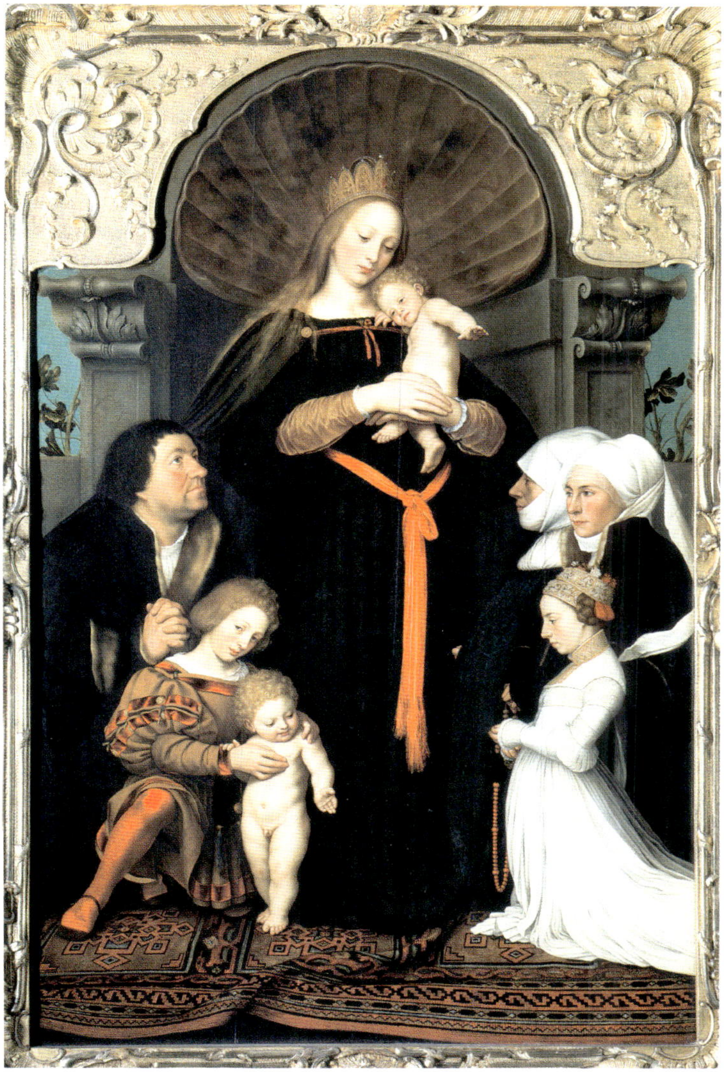

Figure 1. ***Dresden Madonna,***
Bartholomäus Sarburgh,
copy after Hans Holbein the Younger,
Virgin and Child with Burgomaster Jakob Meyer zum Hasen and his Family,
1635–1637

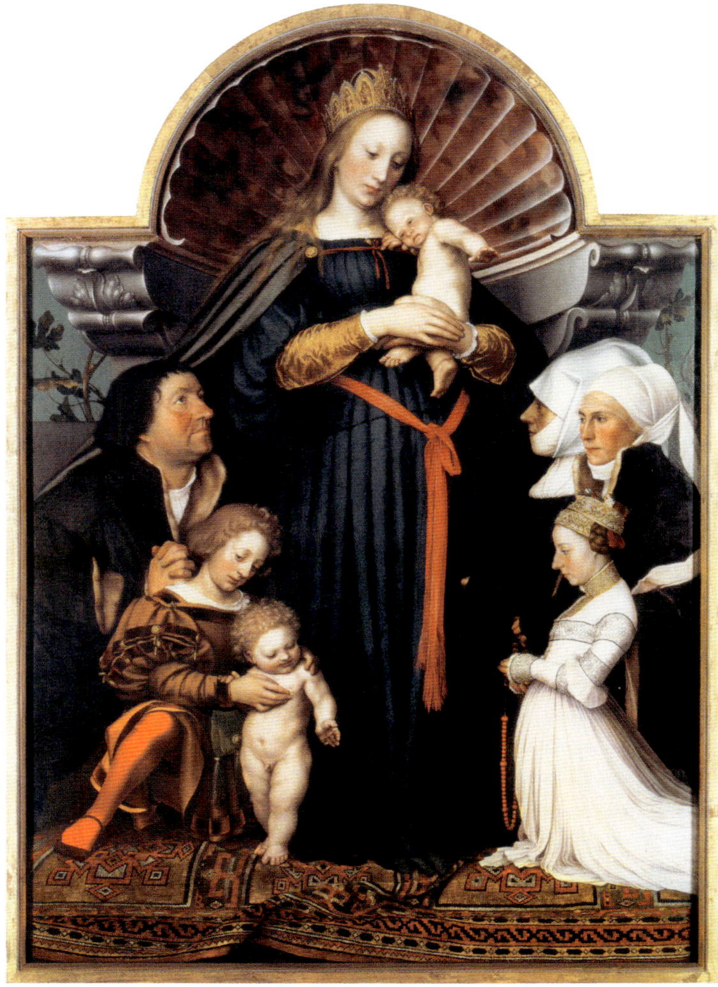

Figure 2. ***Darmstadt Madonna*****,**
Hans Holbein the Younger,
Virgin and Child with Burgomaster Jakob Meyer zum Hasen and his Family**,**
1526/1528

The debate on questions of originals and reproductions, beauty and the power of judgement, reached its climax in 1871 in an exhibition where both works were shown together.

At this exhibition, Gustav Theodor Fechner ventured to conduct a very special experiment. In his *Elemente der Psychophysik* (Elements of Psychophysics) published in 1860, he had already attempted to describe an »exact science of the functional or dependent relations of body and soul or, more generally, of the material and the mental, of the physical and psychological worlds«, in other words, to establish a theory of how external stimuli affect a person's inner experiences.[4] At the exhibition, he then laid out an ›Album‹ (today one would no doubt call it a ›Visitors' Book‹) where he asked visitors to »comparatively evaluate the two main paintings in the exhibition« (Fechner, *Bericht*, p. 1).[5] In doing so, Fechner deliberately set his research apart from the »question of beauty, of interpretation and of the artistic technique of the two major paintings to be compared« (ibid., p. 6) since this requires »a thorough, specialist investigation« (ibid.). Quite in accordance with his field of interest – the experience of beauty – Fechner was above all interested in knowing »1) Which of the two paintings, disregarding its overall connection to art history, considered as a totality, [and] 2) Which of the two heads of the Virgin Mary, in particular, make such a more advantageous, more appealing impression on the beholder that he would rather wish to give the one or other a place in his rooms to view it constantly or repeatedly« (ibid.). In brief, then, he wanted to find out which of these two paintings visitors to the exhibition preferred. Fechner's plan for establishing which picture the beholders liked more, which was *more pleasing,* and also *how* he approached this question, via empirical research using a survey, was to become an important milestone in the history of science.

Adolf Zeising – An Anachronist?

Fechner's research was conducted in the nineteenth century, a period characterised by »an entire series of formalisation, quantification and datification processes.«[6] Advancing industrialisation and mechanisation led (among other things) to re-measuring the human body in order, for example, to produce industrially manufactured clothes in series. Although the endeavour of measuring the body may have been nothing new, there was a radical methodological shift. Whereas previously the human body had been primarily defined by a system of proportions and ratios, the approaches now tended to employ »statistical and arithmetic«[7] methods to obtain, as far as possible, universally valid (and hence highly utilisable) means and averages.

In 1854, against this background, Adolf Zeising published his *Neue Lehre von den Proportionen des menschlichen Körpers, aus einem bisher unerkannt gebliebenen die ganze Natur und Kunst durchdringenden morphologischen Grundgesetze* (New Theory of Proportions of the Human Body, developed from a previously unrecognised morphological law which permeates all of nature and art). In this context, his book was quite remarkable, since in contrast to the efforts of his contemporaries to gather empirical data, Zeising regarded the crucial factor as »the fundamental principle of all creation striving for beauty and totality in the realm of nature just as it exists in the field of art, and that this principle was present as the highest goal and ideal from the very first beginning for the creation of all forms and formal relations, the cosmic as much as individualising, the organic as much as the inorganic, the acoustic as much as the optical, although it has only found its most perfect realisation in human form«.[8] Zeising's *fundamental principle* here is the golden ratio. Admittedly, his approach was not to derive statistical averages and means from his studies. Instead, he was primarily concerned with the question of *beauty* – and in his view, beauty first and foremost related to proportions: »The notion of proportionality is most intimately connected with the idea of beauty. [...] The highest beauty is that which combines, in perfect harmony, all those qualities playing a part in creating beauty« (Zeising, *Neue Lehre*, p. 131).

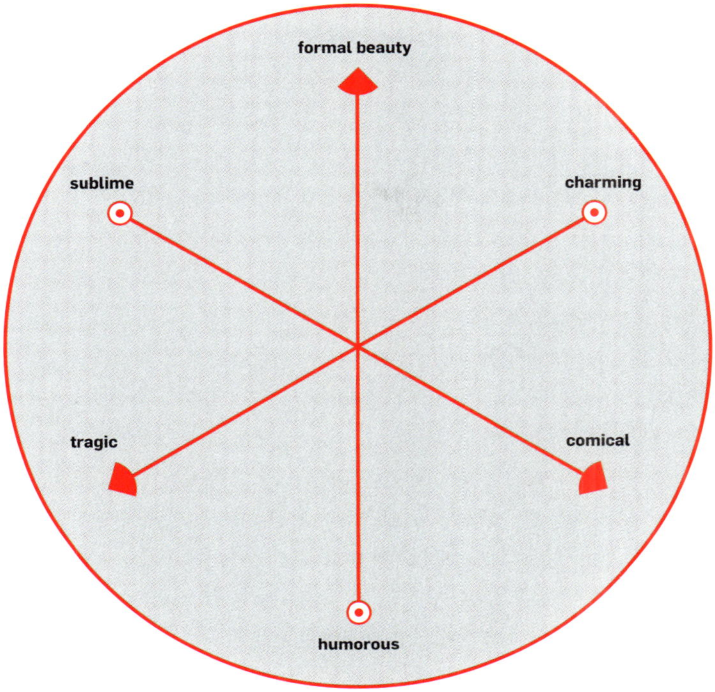

Figure 3. »Circle of Beauty«, Zeising, 1854

With his *Neue Lehre*, Zeising, who was a professor of philosophy in Munich,[9] believed he was casting light on a blind spot in philosophical research into the beautiful. As he noted (Zeising, *Neue Lehre*, p. 3):

»While, on the one hand, natural scientists, namely, physiologists and anatomists and, on the other hand, the practical artists [...] have always addressed this question with a keen interest [...] and even in the more and most recent times have made strenuous efforts to submit this object to ever new investigations: curiously, the philosophical aesthetes particularly, at least those of the modern age, have proven to have a marked indifference towards this question, which is such a cardinal issue for them, indeed regard it as something unimportant and fruitless and have expressly refused to discuss it.«

At the same time, Zeising was clearly aware of the empirical scientific approach taken by his contemporaries, since his introduction explicitly mentions the works of the sculptor Johann Gottfried Schadow, who sought to »express the diversity of the human body via metric proportional descriptions« and »canonically establish his concrete, comparable measurement data as an ideal of the ›average figure‹«.[10] Similarly, Zeising also mentions the statistician Lambert Adolphe Jacques Quetelet, who not only used his own research, but also data from the military on age, height, weight and fertility and traits such as »courage or criminality«, to scientifically determine the features of the *homme moyen*, the ›average man‹.[11]

The new theory of the proportions of the human body – A fundamental law?

Zeising structures his *Neue Lehre* in two parts. In his »Historical Overview of the Previous Systems« (Zeising, *Neue Lehre*, Overview of the Contents), he begins with the classical Greek philosophers Pythagoras and Plato. In the central section, he then deals with the »practical artists, anatomists and physiologists« (ibid.) including, among others, Leonardo da Vinci and Albrecht Dürer, before turning to the »newer philosophers« (ibid.) such as

Kant and Hegel. In his discussion of Kant, in particular, Zeising complains of a »flaw« (ibid., p. 117), and neglecting beauty's »sensory nature revealing itself in space and time« (ibid.): »Therefore, Kant does not offer anything for our question which would be of substantial advantage for the same, indeed some of his remarks, for example that he grants crystals, regions, flowers and similar objects a free and independent, and therefore a higher beauty while horses, people and so on are only judged to have a dependent and, for this reason, a lesser beauty, show that, despite the astuteness with which he dissects the beautiful, he has not been especially successful in this particular field« (ibid.). However, this is an interpretation which may well fail to do full justice to Kant's logic. Nonetheless, here Zeising leaves no doubt regarding his claim to be dealing with the purported blind spot in philosophy, and his rightful place in the group of major philosophers since antiquity – as, after all, he is the one who is confronting and trying to solve this question.

In his book, Zeising then turns to his *Systematic Section,* which also comes with the heading *Development of my own System.* Here, he starts with a *Modification of Beauty,* a definition of terms which leads him to the schematic diagram entitled the »Circle [...] of Beauty« (ibid., p. 144; Fig. 3, p. 135) with formal beauty *(Formell-Schön)* set opposite the humorous *(Humoristisch)* the tragic *(Tragisch)* juxtaposed with the charming *(Reizend)* and the sublime *(Erhaben)* opposite the comical *(Komisch).*

Yet only *formal beauty* alone (something which has a *beautiful form*), he argues, can have a »direct connection with proportionality« (ibid., p. 146): »Form namely endows an outward appearance with the stamp of perfection precisely by most perfectly delineating it from everything else and completing it in itself to such an extent that within these limits the beholding subject forgets everything lying outside and believes he has before him, in this delineated phenomenon, the finite the infinite, the all and the one, the perfect per se and all that exists« (ibid., p. 143). This is a reasoning which, quite in Zeising's sense, presages great things.

His aspirations were similarly high when it came to defining the role of the proportions, since his declared goal was nothing less that finding »a law of proportion which is to be truly satisfying, [and which is to be] inmost and indispensably related to the universal laws of beauty, just as much as to the individual beautiful phenomena« (ibid., p. 158) and which is »that level of formal beauty that dissolves the opposites of the unit and the infinite, of sameness and difference, to achieve harmony by, starting with the bisection, subdividing the totality originally conceived of as a unity into two unequal parts, but giving these parts such a just proportion that, on the other hand, the disparity of the parts is equalised by the parity of the relationship between the totality and its parts« (ibid.). It was this ambitious aim which led him to formulate a »law of proportions in compliance with this notion« (ibid.). In his view (ibid., p. 159):

»When the subdivision or construction of a totality in unequal sections is to appear proportional: then the smaller part must bear the same proportion to the larger part, irrespective of its just measure, as the larger to the totality; or in the reverse arrangement: the totality must have the same ratio to the larger part as the larger part to the smaller.«

As far as Zeising was concerned, this was *the* »aesthetic or balanced proportion« (ibid., p. 162) and subsequently, possibly also due to the way it was described both in terms of language and content, was repeatedly referenced, discussed and mystified. But Zeising did not stop at a purely philosophical definition. Without citing any particular source, he wrote, »The mathematicians call the division of a given line discussed here, the ›division into extreme and mean ratios‹ or the ›golden ratio‹« (ibid., p. 163). Here, then, he introduced the golden ratio as a term, but not without seeking to enlist support for his writings from the hard and hence, verifiable, sciences. He continued, »I do not know the reason for the latter title; but it most likely is due to the fact that one had the proper eye to recognise the exceptional advantages of this ratio which one gains through this division, and the perfection of the proportions which this ratio forms« (ibid.).

He then moved on to a number of mathematical calculations intended to consolidate his *Neue Lehre.* »Hence, even from a purely mathematical perspective,« he notes, »one cannot fail to recognise the perfection of this ratio, and the proportions and progressions based on it [...]« (ibid., p. 173). In his further considerations, Zeising then intensively applies his *law of proportion* to human morphology, before turning to its cosmological, geological, biological (plants, animals, cells, etc.) and architectural applications, and ultimately to the arts and humanities – music, poetry, logic and moral philosophy. Zeising closes his work (ibid., p. 450) with an unmistakably forthright anthropocentric view and the almost prophetic words:

»Wherever [this] gaze turns, whether up to the starry heavens, the image of infinity, or down to the ground of our limited existence, to the spheres of the elements or the realm of individuals, to the fields of inorganic or organic, natural or artificial creations, to the world of plastic or tonic phenomena, to the forms of poetry or science, to the phases of moral and political or religious life – everywhere we find that the highest degree of satisfaction is derived from those appearances that, more or less, harmonise with the Ur-type of the human form developed here [...], and that even the most determined scepticism could never accept that this is just a mere coincidence: for such a presumption would indeed be blasphemy«.

Who could say anything against such a conclusion – and, in doing so, expose themselves to the possible accusation of blasphemy? And so Zeising's work fuelled a number of debates among his contemporaries.

Fechner's aesthetics from below

Gustav Theodor Fechner also addressed Zeising's work. His *Zur experimentalen Ästhetik* (On Experimental Aesthetics) was published in 1871 – the same year in which he conducted his research into the visual reception of the two Virgin Mary paintings. Regarding aesthetics as a »branch of external psychophysics« (Fechner, *Zur experimentalen Aesthetik*, p. 557), he wanted to find » for [this] field which up until now has seemed to elude any exact investigation, the weak points and the beginnings of such research to enable it to enter the sphere of experiment, measurement and calculation« (ibid.). With the help of such an experiment, his objective was to convert aesthetics, »a theory previously almost solely a matter for philosophical speculation and artistic insights [...]« (ibid.), from a purely theoretical approach into a concrete, replicable empiricism. To achieve this, Fechner referred to Zeising's investigations into the golden ratio, though with some reservations: »From the very start, I admit [...] that I approached the [experimental] research with the greatest want of confidence in the golden ratio, doubting that it has any aesthetic significance at all« (ibid., p. 577).

Just as in the case of the Virgin Mary paintings, Fechner wanted to investigate an aesthetic preference for one stimulus or another (in this case, in the form of a shape). For his experiment, he prepared rectangles with different aspect ratios, including the golden ratio, asking test subjects to select the shape they found most pleasing. In fact, his experiment did show a slightly stronger preference for the golden ratio (approximately 35 % selected this shape from the ten different rectangles).[12]

1 Andreas Beyer, »Am Anfang war der Streit. Hans Holbein d.J. und die kunsthistorische Tradition«, in: Christian Scholl/Sandra Richter/Oliver Huck (eds), Konzert und Konkurrenz. Die Künste und ihre Wissenschaften im 19. Jahrhundert, Göttingen, Universitätsverlag Göttingen 2010, pp. 201–217, here p. 201.
2 Gustav Theodor Fechner, Ueber die Aechtheitsfrage der Holbein'schen Madonna, Leipzig, Breitkopf & Härtel 1871, p. III.
3 A. Beyer, »Am Anfang war der Streit«, p. 201.
4 Gustav Theodor Fechner, Elemente der Psychophysik, Leipzig, Breitkopf & Härtel 1860.
5 Gustav Theodor Fechner, Bericht über das auf der Dresdner Holbein-Ausstellung ausgelegte Album, Leipzig, Breitkopf & Härtel 1872, p. 1 ff; subsequent references noted in the text.
6 Daniela Döring, Zeugende Zahlen. Mittelmaß und Durchschnittstypen in Proportion, Statistik und Konfektion, Berlin, Kadmos 2011, p. 15.
7 D. Döring, Zeugende Zahlen, p. 16.
8 Adolf Zeising, Neue Lehre von den Proportionen des menschlichen Körpers, aus einem bisher unerkannt gebliebenen die ganze Natur und Kunst durchdringenden morphologischen Grundgesetze, Leipzig, Rudolph Weigel 1854, V; subsequent references noted in the text.
9 Gustav Theodor Fechner, Zur experimentalen Aesthetik. Erster Theil des IX Bandes der Abhandlungen der mathematisch-physischen Classe der Königl. Sächsischen Gesellschaft der Wissenschaften N 6, Leipzig, Hirzel 1871, p. 567; subsequent references noted in the text.
10 D. Döring, Zeugende Zahlen, p. 18.
11 D. Döring, Zeugende Zahlen, p. 18.
12 Gustav Theodor Fechner, Vorschule der Aesthetik, Leipzig, Breitkopf & Härtel 1876, p. 195; subsequent references noted in the text.
13 Marie Guthmüller, Wolfgang Klein, »Introduction«, in: Marie Guthmüller, Wolfgang Klein (eds) Ästhetik von unten – Empirie und ästhetisches Wissen, Tübingen, Attempto Verlag 2006, pp. 1–20, here p. 4.

But the really important aspect to Fechner's experiments was actually his differentiation between an aesthetics ›from above‹ and ›from below‹, as he explained in his *Vorschule der Aesthetik* (Introduction to Aesthetics) from 1876 (Fechner, *Vorschule*, p. 1). Here, Fechner clearly differentiated the two concepts by opening up a dichotomy between the experiment and philosophical reflection: »The other approach, in contrast, the approach from below, allows or at least promises a direct clear orientation not only in the field of concepts which is subordinate to the sphere of pleasure and displeasure, but also to the reasons for that pleasure and displeasure [...]« (ibid., pp. 2–3). In his book, Fechner returned to his previous work where – as in the case of the two Virgin Mary paintings – he addressed the impact of art or, as in Zeising's writings, referred to the golden ratio. The book became the »catalyst for the empirical aesthetic in its experimental version«[13] and, hence, for modern perceptual psychology as well. Even if Fechner, given his research interests, might have found other objects of investigation, Zeising's work and, in particular, his claim that the golden ratio represented the law of proportion offered an opposing view which, as the nineteenth century drew to a close, only stands too aptly for the antithesis between verifiability in the natural sciences and theory in the humanities.

Bader, Lena: Bild-Prozesse im 19. Jahrhundert. Der Holbein-Streit und die Ursprünge der Kunstgeschichte, Munich, Wilhelm Fink 2013.
Döring, Daniela: Zeugende Zahlen. Mittelmaß und Durchschnittstypen in Proportion, Statistik und Konfektion, Berlin, Kadmos 2011.
Fechner, Gustav Theodor: Elemente der Psychophysik, Leipzig, Breitkopf & Härtel 1860.
Fechner, Gustav Theodor: Ueber die Aechtheitsfrage der Holbein'schen Madonna, Leipzig, Breitkopf & Härtel 1871.
Fechner, Gustav Theodor: Bericht über das auf der Dresdner Holbein-Ausstellung ausgelegte Album, Leipzig, Breitkopf & Härtel 1872.
Fechner, Gustav Theodor: Zur experimentalen Aesthetik. Erster Theil des IX Bandes der Abhandlungen der mathematisch-physischen Classe der Königl. Sächsischen Gesellschaft der Wissenschaften N 6, Leipzig, Hirzel 1871.
Fechner, Gustav Theodor: Vorschule der Aesthetik, Leipzig, Breitkopf & Härtel 1876.
Guthmüller, Marie/Klein, Wolfgang (eds): Ästhetik von unten – Empirie und ästhetisches Wissen, Tübingen, Attempto Verlag 2006.
Scholl, Christian/Richter, Sandra/Huck, Oliver (eds): Konzert und Konkurrenz. Die Künste und ihre Wissenschaften im 19. Jahrhundert, Göttingen, Universitätsverlag Göttingen 2010, pp. 201–217.
Müller-Tamm, Jutta/Schmidgen, Henning/Wilke, Tobias (eds): Gefühl und Genauigkeit. Empirische Ästhetik um 1900, Munich, Fink 2014.
Zeising, Adolf: Neue Lehre von den Proportionen des menschlichen Körpers, aus einem bisher unerkannt gebliebenen die ganze Natur und Kunst durchdringenden morphologischen Grundgesetze, Leipzig, Rudolph Weigel 1854.

Human dimensions
Tatami mats

Florian Schütz

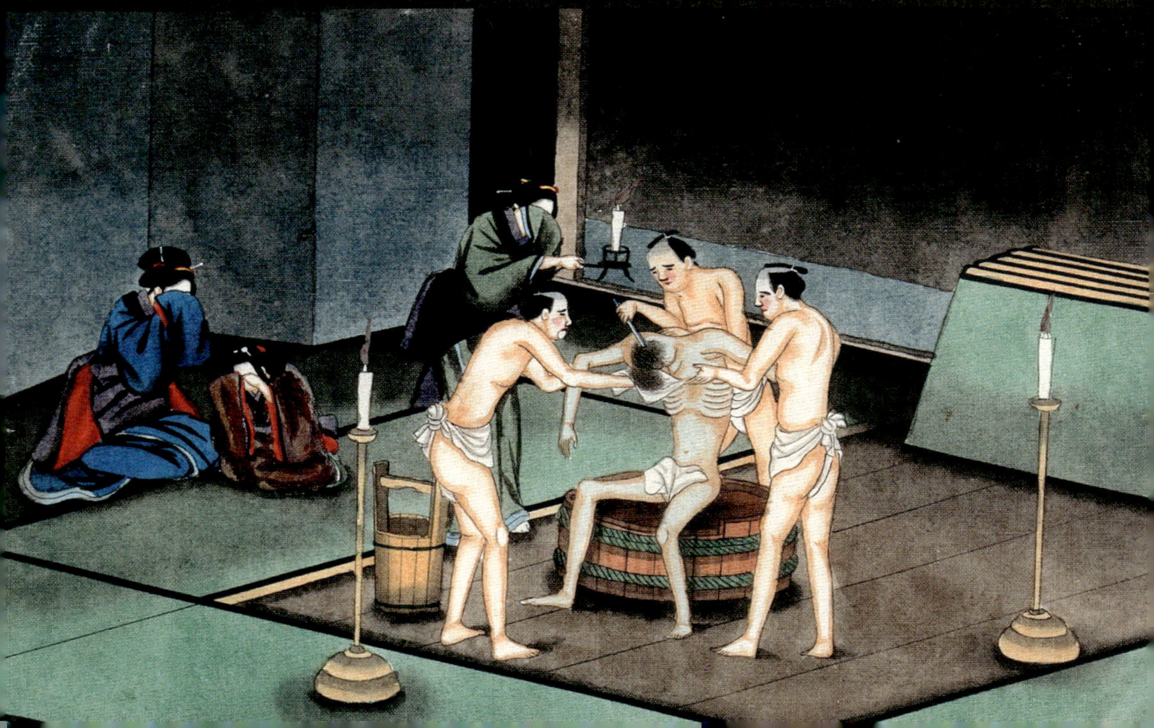

In the 1630s, the Tokugawa shogunate issued a series of decrees creating a ›closed country‹, sealing the Japanese islands off from the ›western‹ world. This policy of national seclusion was fuelled by a growing distrust of the intentions of the Portuguese and Spanish, the *Nambanjin* (›southern barbarians‹). However, the country's total isolation was never realistically feasible. There continued to be transfer processes between Japan and the Dutch East India Company, which was bringing literature and knowledge into this closed country. In the end, *Rangaku* (›Dutch studies‹), as it was called, though never exactly reflecting reality, proved to be a preparation for the country's partially forced opening after the Meiji Restoration in 1868. Although the subsequent modernisation and Westernisation of Japan certainly produced cultural hybrids – such as the western-style clothes adopted in the administration and at official events – it did not destroy the country's rich traditions and customs.

One aspect of the continuation of these customs is evident in the design and measurement of interior space. Traditionally, the area of a room in Japan is given in *tatami* (畳). A *tatami* is a mat made of rice straw used as a floor covering. The mats have a thick rice straw core, a soft rush or reed cover and cotton brocade borders. Since these materials are all sensitive to wear and tear, they should only be walked on either in socks or in bare feet. Despite this, *tatami* mats are not only used in the home, but also in sport venues – for example, by practitioners of judo. The visible signs of wear and tear as well as the aging of the mats are actually desirable, since they show the traces of life in the rooms and make their history perceptible. However, *tatami* mats are most strongly associated with the traditional Japanese room, a *washitsu* (和室). In such rooms, the rectangular shape of the sliding doors and walls covered with lightly opaque paper vertically echoes the form of the *tatami* mats on the floor. Initially, this style of interior, which often has a ceremonial character, was the preserve of the nobility. Although the homes of the lower classes were far simpler, often with just earth floors, the popular *tatami* mats also became common here in the late seventeenth century.

The size of a *tatami* and its *ken* (間), the length of its sides, varies from region to region. Nonetheless, the mats usually cover approximately 1.64 m^2, the space needed to accommodate two men sitting facing each other, with the sides following a 2:1 ratio. For example, in Kyoto, a *tatami* measures 0.955 m x 1.91 m (*Kyoma*, 京間), in the region of Nagoya 0.91 m x 1.82 m (*Ainoma*, 合の間) and in the Tokyo area 0.88 m x 1.76 m (*Edoma*, 江戸間 or *Kantoma*, 関東間). The thickness of the *tatami* also varies between 5 and 6 cm.

Possible arrangement of 4 $^1/_2$ mats

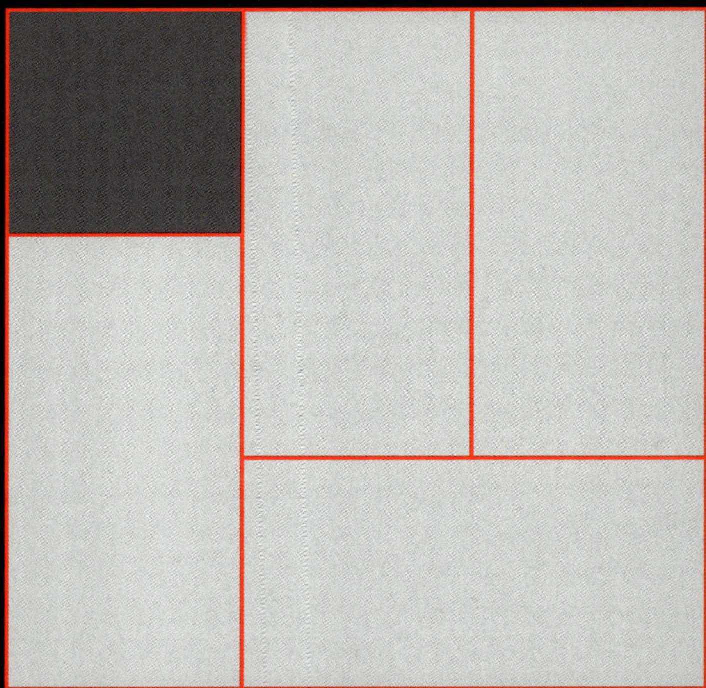

Since *tatami* offer a convenient way of estimating the scale of an apartment or house, real estate advertisements include the number of *tatami* fitting in the space to indicate the size of the property. Although the *tatami* could be combined in many different ways, the patterns often follow rules prescribed by the purpose of a room. A standard room for the tea ceremony, for instance, often measures 4 $^1/_2$ mats, a rectangle with three mats set in a T-shape and the other mats laid alongside. A half mat is known as a *Hanjo* (半畳). Since this marks the location of the brazier or sunken hearth used on cold days, it is set in the centre of the other mats. The layout of the mats also reflects the order in which guests sit, and hence their status or role. The host and guests use different entrances to the tea room and sit in different places. The layout of the mats can be either auspicious or inauspicious, and wrongly laid *tatami* are thought to bring bad luck for the people living there.

The *tatami* mat can also be scaled to provide a unit of measurement for larger surfaces such as, for example, the size of a parcel of land. A *tsubo* (坪), which is the size of two *tatami*, is also a standard unit for the size of fields. In contrast to *tatami*, which vary in size regionally, the area of a *tsubo* is standardised throughout Japan as 3.306 m^2.

In the history and architecture of Japan, the *tatami* mat plays a major role in traditional homes and measurements. However, although the metric system (i.e. including square metres) was adopted a long time ago in Japan and overall measurements of new apartments or homes are less commonly found only in *tatami* today, formal or social festivities are still usually celebrated in a *washitsu*.

The battle of proportions
The Fechner Experiment

Holger Höge

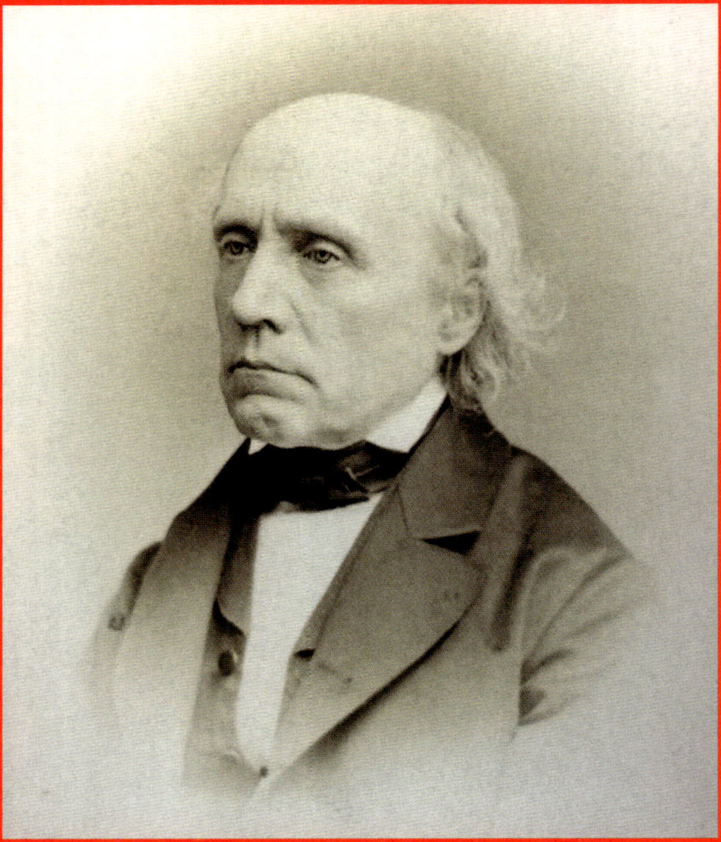

Gustav Theodor Fechner (1801–1887), c. 1870

Nearly all the objects we encounter in daily life are comprehensible in terms of measure and number. Since they are three-dimensional (width, depth, height), the extension of each dimension is always set in relation to the others; the objects are proportioned. This applies just as much to windows and doors, for example, as it does to bars of chocolate, admission tickets, house façades, pictures, etc. Particular attention is often paid to art works and photographs when the aim is to ensure that they have the right, i.e. pleasing, proportions (for photos, for instance, 9 x 13 cm). These formats are produced in proportions intended to meet specific criteria, and this is so much a part of daily life that we hardly ever think about it. It only becomes more complicated if we question how those *specific criteria* are to be understood.

One of the possible criteria here could be, for instance, *ease of handling*. The common paper format in Germany (the DIN A standard) has the major advantage that every DIN A standardised sheet can be transformed into the next larger or smaller format by doubling or halving the paper surface. Nearly everyone is familiar with the standardised A4 (210 x 297 mm) format. This fits all the major commercial printers, for example, and since it is also the size of all letters from authorities, it has in a sense become the official format.

But what happens when the criterion is changed? *Ease of handling* may be appropriate and sensible for printers or authorities, but what happens if we take beauty as the key criterion? As a rule, we prefer to surround ourselves with well-designed objects, rather than unsightly or even ugly ones. So the appearance of such objects should have *pleasing proportions*, and therefore the objects meeting such a criterion ought to be preferred over others. From this perspective, then, we would select from the mass of available objects those whose proportions best fulfil some criterion of beauty – from personal appeal to elegance, sublimity, positive impression, and so on.

With more knowledge of this process, one would thus also be in a position to establish a rule for the proportions of objects – and architects, product designers, artisans or artists would have a tool available allowing them to give many – if not all – objects in our surroundings an appealing and pleasing, that is *beautiful*, appearance. Such an idea is not new. It was already a subject of discussion in the nineteenth century, and is linked to the name and work of Adolf Zeising.[1] In his book *Aesthetische Forschungen* (Aesthetic Research) published in 1855, Zeising collected numerous instances of objects whose pleasing appearance could be defined by the proportions of the golden ratio. Although the golden ratio had been known since Luca Pacioli (c. 1445–1517), Zeising endowed it with the status of a universally valid theory of proportions. Zeising's book was very influential and reached a wide audience, including Gustav Theodor Fechner (1801–1887), one of the founding figures in psychology. But Fechner doubted whether these claims were true. Did the golden ratio really always create a pleasing appearance, wherever it was applied? Was it right to elevate it to a universal law of aesthetics? Until the mid-nineteenth century, *beauty* was considered the main criterion for art – and it was in this context that a fierce debate erupted over the validity of the golden ratio's pleasing appearance.

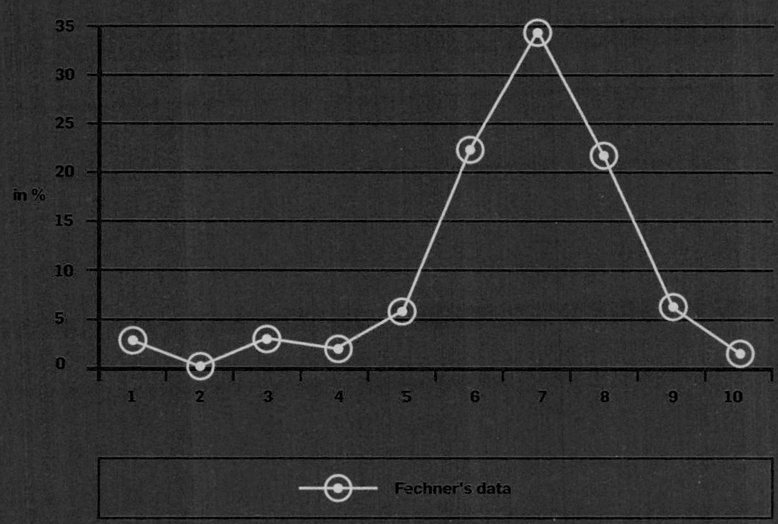

Putting the Fechner experiment to the test

In accordance with the spirit of a sceptical, empirical scientist, Fechner wanted to test this claim experimentally. He started from the hypothesis that if the golden ratio really does result in aesthetically pleasing objects then, first, this must also apply to simple objects and, second, the maximum preference for the golden ratio's proportions would be most pronounced if a person could choose between several differently proportioned objects – and this was exactly the subject of his investigations in an experiment that was to become famous.[2]

Fechner took ten four-sided shapes with aspect ratios representing a variety of proportions (with the same surface area, since then Fechner could exclude, for example, a choice of some particular shape as most pleasing due to its surface area). The four-sided shapes included a square (aspect ratio 1:1) and the golden ratio (aspect ratio 1:0.681). In the graph (left), the four-sided shapes, each assigned a number from 1 to 10, are plotted on the X axis, whereby No. 1 is the square and No. 7 the golden ratio. Over a period of several years, Fechner asked nearly 400 people to take part in his experiment, showed them the shapes and asked them to choose the one they considered most aesthetically pleasing.

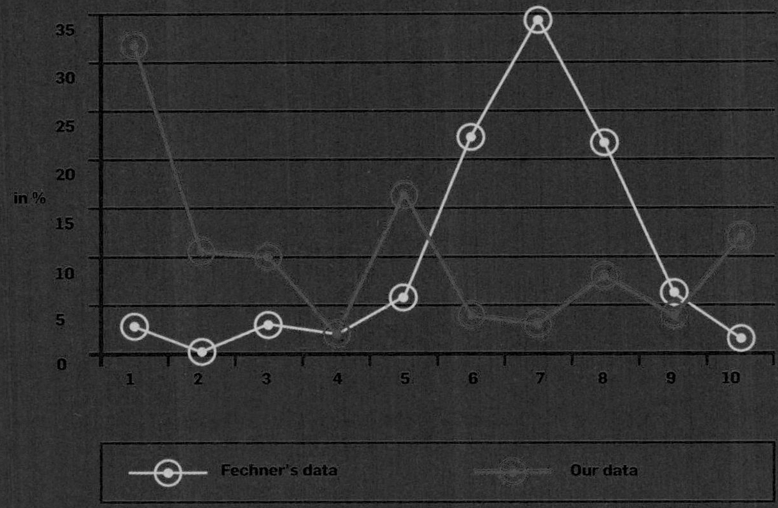

1 Adolf Zeising, Aesthetische Forschungen, Frankfurt/Main, Meidinger 1855.
2 Gustav Theodor Fechner, Vorschule der Aesthetik, Leipzig, Breitkopf & Härtel 1876, pp. 184–202.
3 Holger Höge, Emotionale Grundlagen ästhetischen Urteilens, Frankfurt/Main, Peter Lang 1984.

The golden ratio shape (No. 7) was most frequently chosen and so represented the maximum preference, just as Zeising's theory had predicted. At this point, one could close this chapter with the universal law of aesthetics confirmed. But unfortunately, it is not quite that simple.

For a principle to have the status of a *universal law*, we rightly expect it to be valid always and in all circumstances. Consequently, the preference for the golden ratio must also remain stable over time – and therefore, Fechner's nineteenth-century findings ought to be confirmed by any similar research conducted in the late twentieth century as well. To test this, we replicated the same experiment in the late twentieth century – similarly conducted over a number of years – using four-sided figures with the same proportions.[3] However, as the graph shows (above), we obtained very different results.

In our experiment, the square was considered to be the most aesthetically pleasing shape. Since the discrepancy between the two experiments is considerable, one might well assume that some error had slipped into the experimental process unnoticed and was responsible for this difference. To exclude this possibility, we repeated the experiment several times, but always obtained the same result: the square was the most preferred shape! So how can these findings be explained? On the basis of everything we know about Fechner's experiment, one cannot exclude the possibility that the participants knew about the golden ratio. He questioned members of Leipzig's arts association (*Leipziger Kunstverein*), making sure that his test subjects came »from the most diverse, though always educated, classes« (Fechner 1876, p. 194). And precisely this group of individuals could very well have read about a ›universal law‹ in Zeising's writings – and therefore selected the four-sided shape No. 7. Unfortunately, such an assertion can never be proven since the documents detailing Fechner's experiment have not survived. Moreover, it is doubtful whether he collected any of his test subjects' personal details at all. The benchmarks for properly documenting an experiment – according to today's standards – had not yet been developed.

The reality of life changes – and tastes as well?

Our findings can at least unequivocally show that there is no evidence to support Zeising's claim that the aesthetic pleasure of the golden ratio is universally valid. From this, it also follows that we cannot expect the findings of our experiment to be repeated in exactly the same way in 100 or 200 years' time. Hence, the question that remains is: Why did the test subjects behave differently in 1876 in comparison to 140 years later? Since in 2016, for obvious reasons, we are unable to ask the 1876 sample directly, there can be no definite answer. Instead, we can only posit reasons why, such as, for example: the lifeworld in 1876 is certainly not identical with the situation today; everything (from politics to fashion to transport, etc.) has changed – and hence people's tastes may also have changed. In itself, this does not solve the problem, since such an assertion would make it necessary to conduct research to test the ›hypothesis of changing tastes‹. So the last word on this issue has yet to be spoken!

Höge, Holger: »Fechner's experimental aesthetics and the golden section hypothesis today«, in: Empirical Studies of the Arts 13 (1995), Los Angeles et al., SAGE Publications, pp. 131–148.

Höge, Holger: »The golden section hypothesis – A funeral, but not the last one«, in: Visual Arts Research 22 (1996), Champaign, Illinois University Press, pp. 79–89.

Höge, Holger: »Why a special issue on the golden section hypothesis? An Introduction«, in: Empirical Studies of the Arts 15 (1997), Los Angeles et al., SAGE Publications, pp. 111–114.

Höge, Holger: »The golden section hypothesis – Its last funeral«, in: Empirical Studies of the Arts 15 (1997), Los Angeles et al., SAGE Publications, pp. 233–255.

Höge, Holger: »The minute variations: Paper quality, colour, angles and the preference for proportion«, Bonaiuto, Paolo et al (eds.), Proceedings of the 15th Congress of the International Association of Empirical Aesthetics, Rome, Edizioni Universitare Romane 1998, p. 153.

Höge, Holger: »Keine Schönheit ohne Maß«, in: Zeitschrift für Literaturwissenschaft und Linguistik 152 (2008), Stuttgart, J. B. Metzler, pp. 25–41.

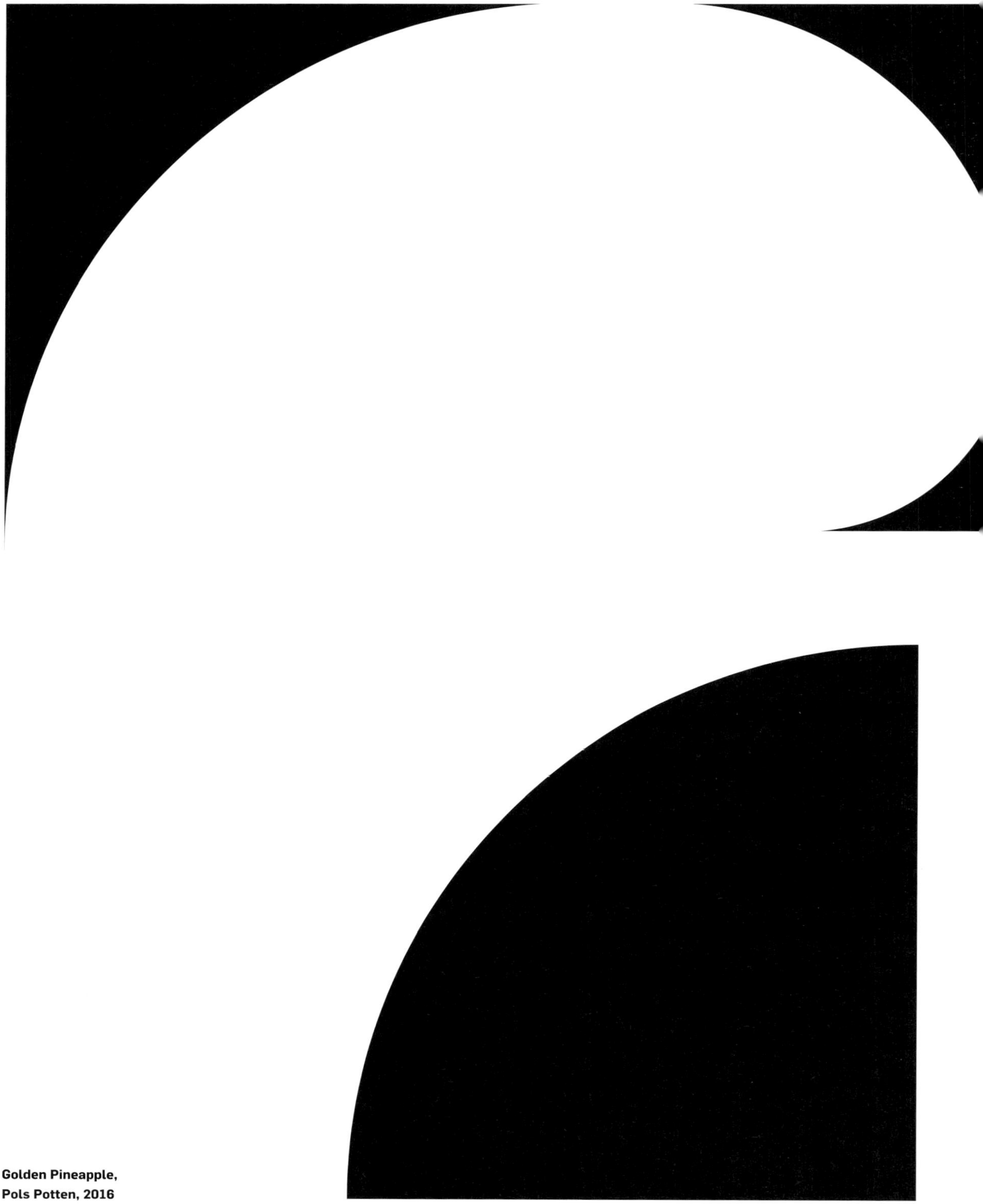

Golden Pineapple,
Pols Potten, 2016

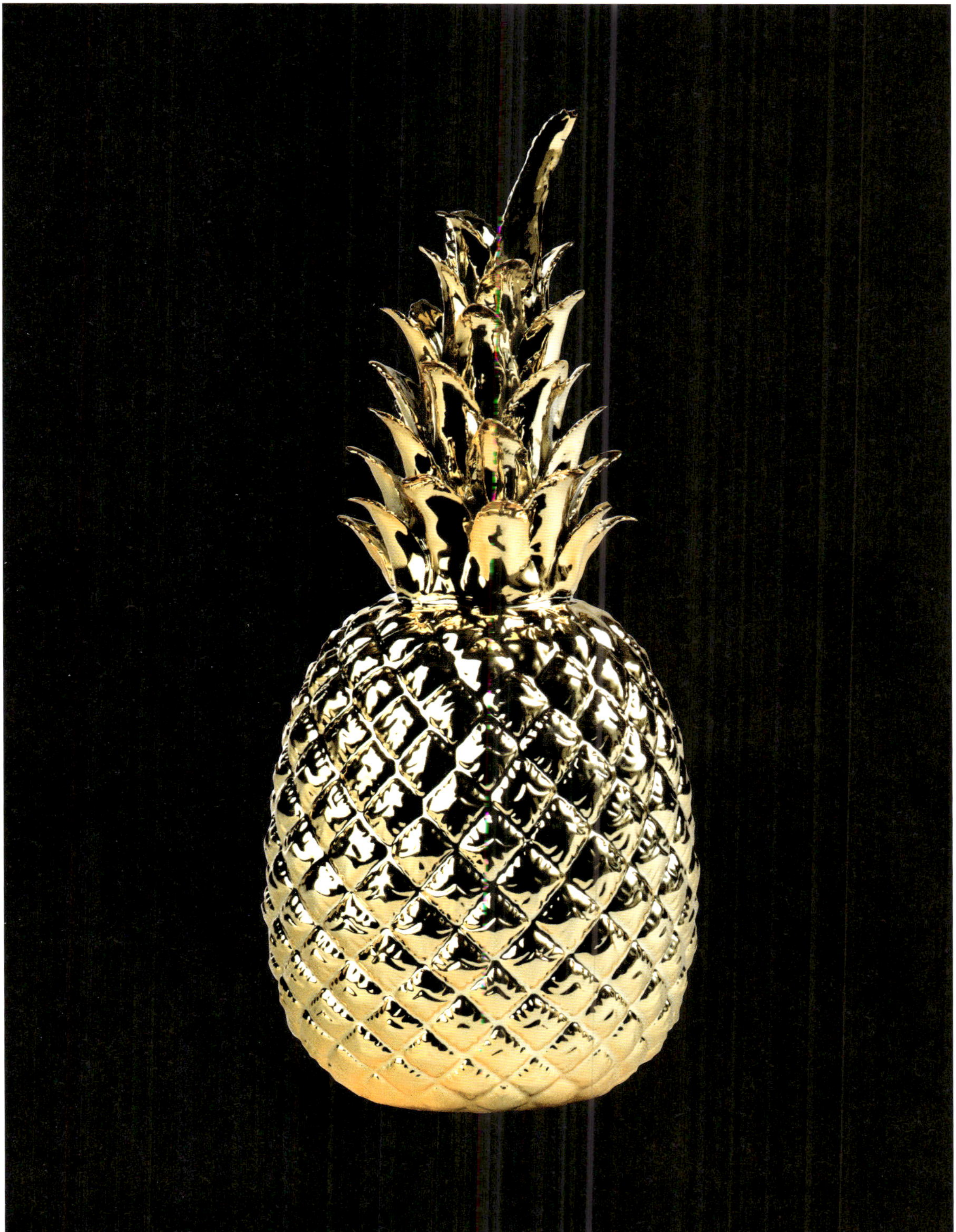

Two offset Fibonacci sequences

A winning formula for organ development in plants

Peter Leins, Claudia Erbar

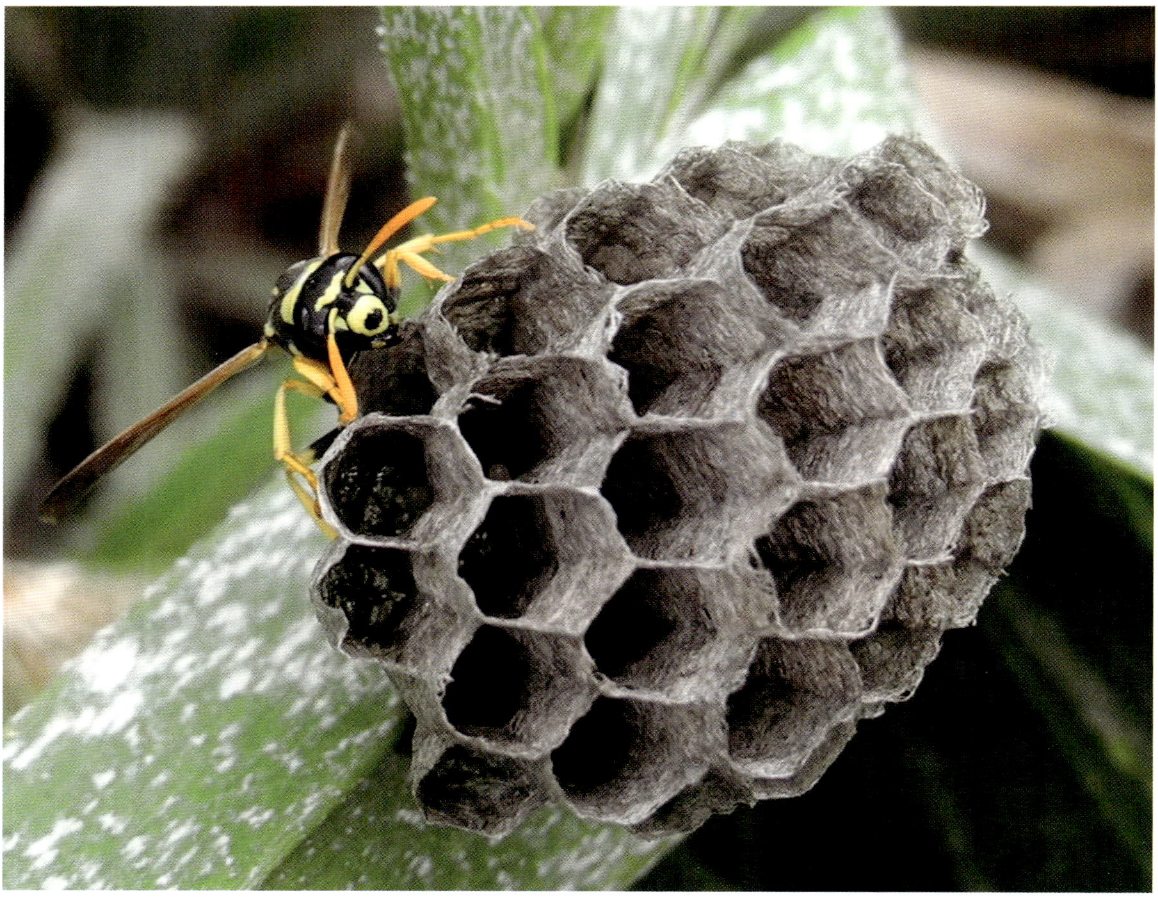

Efficiency is the hallmark of organismic evolution, and optimisation its ruling principle. The goal of optimisation is to achieve the greatest possible efficiency with the least possible effort. Both parameters in this process (*input* and *output*) are variables. Natural selection – an optimisation process through which organisms adopt the most advantageous new structures and functions – is marked by fierce competition. Seen as a canalising mechanism of evolution, it is tantamount to mass extinction. The winners ›get lucky‹, as all the other mechanisms of evolutionary change are random processes. These mechanisms include (1) changes to genetic material (mutations); (2) genetic recombination through sex, the result of two powerful processes in which sex cells are divided (meiosis) and recombined; (3) migration (the spread of genetic diversity); and (4) isolation (geographic, reproductive, or both). This last mechanism can even lead to the development of new species. Everything is thus a matter of chance. Like physical adaptations, successful ›strategies‹ also evolve randomly and come to exert considerable selection pressure in their environments.

One of the great strategies that emerged over the course of plant evolution is the arrangement of leaves, flowers, and their primordia in ways that optimise their functionality. These include the utilisation of light (leaves) and space (organs in flowers and flowers in inflorescences), as well as adaptation to highly complex networks of correlations, for example reproduction and thus evolution itself.

Leaves can be arranged along the stem in such a way that groups of them form at single nodes, with the leaves of successive nodes being ›staggered‹. Multijugate arrangements of this kind, featuring whorls of leaves radiating from a single node, are relatively rare among flowering plants. In the case of the common mare's tail (*Hippuris vulgaris*), the lower parts of which grow under water, a large number of needle-shaped leaves grouped at single nodes serves the overall purpose of enlarging the circumference of the plant, which has a beneficial effect on the supply of oxygen in the water.

Figure 1. Queen wasp in her first hive. The hexagonal shape of the individual cells, arranged in staggered rows, allows optimal space utilisation.

Two other leaf arrangements are much more common: decussate (two leaves at every node, with successive leaf pairs rotated 90 degrees relative to one another) and alternate, also called spiral (one leaf at every node). The leaves of successive nodes are usually at a certain (constant) angle to each other. This is called the divergence angle. Common angles are 180°, 120°, 144°, 135°, and 138.46°.

Using the Fibonacci series to measure divergence angles

Two brilliant scholars, Karl Schimper (1803–1867) and Alexander Braun (1805–1877), were the first to show that divergence angles can easily be determined without using a protractor; all that is needed are two offset Fibonacci sequences (Braun 1835). They created a series of fractions, now called the Schimper-Braun series. This series, like the Fibonacci series, is infinite, and begins with the terms $^1/_2$, $^1/_3$, $^2/_5$, $^3/_8$, $^5/_{13}$, $^8/_{21}$, $^{13}/_{34}$ etc. While the set of numerators contains a complete Fibonacci series, the denominators start with the number 2 (the first two Fibonacci numbers – 1, 1 – are not included). The Schimper-Braun fractions refer to an amount of rotation around a stem. To express a divergence (angle between two successive leaves) in degrees, the respective fraction is simply multiplied by 360.

This can be illustrated with a few brief examples of different divergence angles, starting with the leek which has a divergence angle of 180° ($^1/_2$). Since the young leaf primordia are stem-clasping, successive leaf primordia are located just opposite the previous site, somewhat higher on the stem. In the case of the leek, this divergence of $^1/_2$ is related to a functional characteristic in which the leaves, tightly packed at the plant's short stem, overlap, and every leaf develops a long, closed, cylindrical leaf sheath at its base (the rings seen when a leek is cut in half crosswise). Taken together, this bundle of overlapping leaf sheaths forms a stable ›pseudostem‹ above the stem, which remains short in the first year of growth.

From a culinary perspective, first-year leeks are tender from late summer into the winter and consist only of leaves once the cook cuts off the base. They lose their tenderness later as a long, tough stalk develops in the spring and summer of the following year. Emerging from the short stem, it pushes the developing inflorescences up from the pseudostem.

A divergence angle of 120° ($^1/_3$) can be typical of an entire plant family, for example sedges (*Cyperaceae*). Many examples of a $^2/_5$ divergence (144°) can be found, for instance in the leaf rosettes of the hoary plantain (*Plantago media*), a common plant in areas shaped by human activity. There is variation within the species, however, with some plants having a leaf divergence of $^3/_8$ and some showing an average angle of approximately 137° when the divergence is measured with a protractor. As differences between successive terms in the series become ever smaller, the fractions approach what is called a ›limit divergence‹, although naturally there is no limit literally in reality. Since the fractional series itself never ends, it becomes infinitely closer to this limit – an irrational number with an approximate value of 137.5° – but never actually reaches it. The divergences $^5/_{13}$ (138.46...°), $^8/_{21}$ (137.14...°) and $^{13}/_{34}$ (137.64...°) come so close to the limit that, in practice, any further divergences can be safely ignored.

So who needs a protractor? All we have to do is count successive leaves – either from bottom to top (that is, in the order in which they emerged), or top to bottom (which is usually easier) – until we reach a leaf that is above the first leaf. The last leaf is not counted. The result – assuming that the angle is more or less constant – is a number in the Fibonacci series, more specifically a numerator in a fraction contained in the Schimper-Braun series. The number of turns taken around the stem before reaching the leaf above the first leaf is the number in the denominator. This value can also be found simply by looking at the relevant term in the Schimper-Braun series.

It is fun to experiment with these ratios in the kitchen: the next time you cook a red cabbage (a giant leaf bud), cut it crosswise near the base before you chop it up, and count the very pronounced white central veins of the leaves folded within the bud primordium (Fig. 2).

Now it should be clearer why, assuming we are unaware of the simple counting method, we use a protractor with some hoary plantain plants, since we are dealing with angles that are approaching the limit divergence. This is indicated by the fact that no leaf can be found which is above the first leaf. The limit divergence, in its irrationality, divides the arc of the circle in the golden ratio.

The golden ratio maximises efficiency

The golden ratio coincides with the efficiency rule when there exists a certain correlation between the shoot apex (shoot apical meristem) and the leaves that form on it. This happens quite commonly, and allows the optimal utilisation of space, not to mention the best use of light for photosynthesis whenever leaf size decreases as successive leaves are formed. An optimal ratio between the shoot apical meristem and the leaf primordia is presumably a very old adaptation that evolved many times in the development of the first cormous plants approximately 400 million years ago. These plants, including ferns and later seed plants, have a ›corm‹, that is, a plant structure consisting of the three basic organs: root, stem, and leaf. Apart from horsetails (which have a whorled leaf arrangement), many of the plants in this group have alternate leaf arrangements and some of them may even have reached the medium-range and higher divergences of the Schimper-Braun series, which approach the golden angle.

While it is not an explanation, the inhibition field theory, which posits inhibition fields in the shoot apical meristem, provides a useful model for the optimal formation of leaf primordia, in terms of both position and timing. According to this theory (Schoute 1913), the position of a new leaf primordium is determined by the interplay between two competing inhibition fields of varying strength.

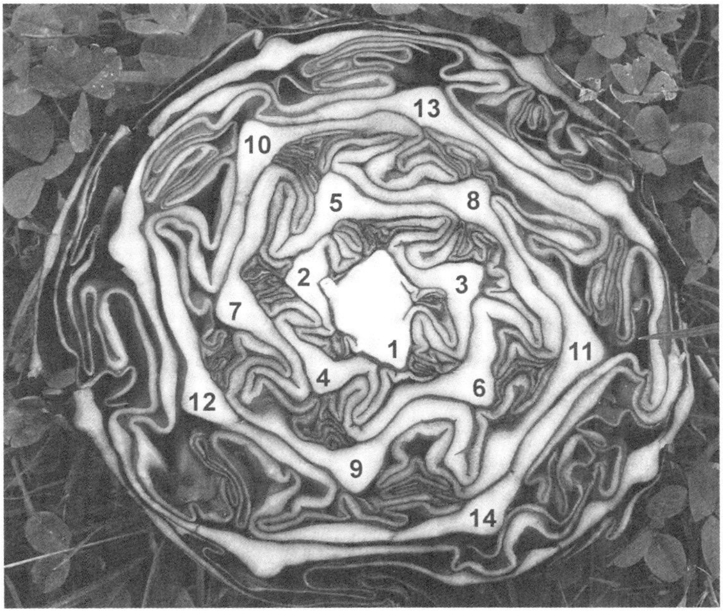

Figure 2. Cross-section of a red cabbage. The numbers on the central veins of the folded leaves show the order of the leaves from inside to outside (that is, the reverse of the order in which they developed). The fourteenth leaf is located above the first, and is not counted. The result is a denominator in the Schimper-Braun series, namely 13. The numerator is therefore 5, and the divergence angle is 138.46°.

In one case, there is an inhibition gradient, which starts at the tip of the shoot apical meristem and progressively loses its inhibitory power the further it travels toward the base, until at last a distance from the tip to the already emerging leaf primordia – referred to as the minimal distance – is reached, at which point a new leaf can form. This accounts for the fact that in multijugate versions of whorled leaf arrangements (a relatively uncommon arrangement in which multiple leaves are grouped around a single node), the leaves of successive nodes alternate. The common mare's tail is an example of such an arrangement. The minimal distance is not achieved until the gradient reaches the gaps between young leaf primordia (Hofmeister's rule, 1868). In another case describing alternate leaf arrangements, the most recently formed primordium is surrounded by an inhibition field which prevents the immediate formation of another primordium directly next to it. As the primordia age, their inhibition fields become progressively weaker.

An inhibition model such as this guarantees – in conjunction with certain size ratios – that primordia form on the shoot apical meristem from the base to the top or from outside to inside, as close to each other as possible.

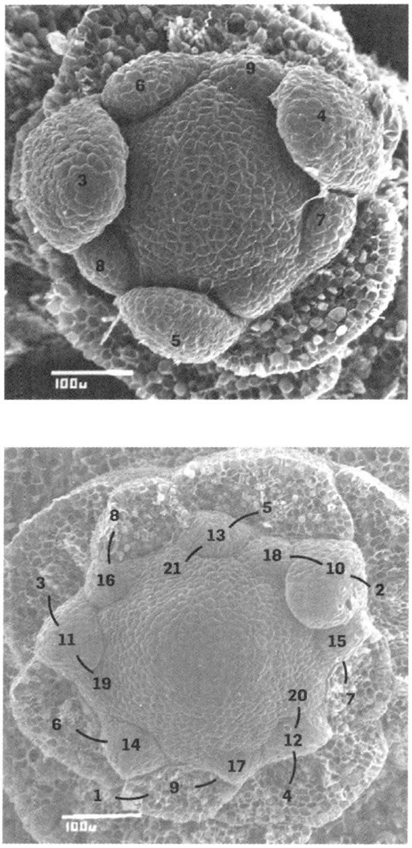

Figure 3. Four successive stages in the development of flower buds of the star anise (*Illicium anisatum*), seen from above through a scanning electron microscope. The numbering indicates the sequence of the primordia in the golden ratio. Eight rows of primordia spiral out from the centre, resulting in a parastichy number of eight (Leins/Erbar 2008, 2010, altered).

Can the golden ratio be observed in flowers?

Most flowers do not present any evidence of the golden ratio, at least at first glance. Even as children, we typically drew flowers with five colourful petals, green sepals, and something undefinable – usually yellow – in the middle. Indeed, many flowers are composed of flower whorls: a whorl of sepals, a whorl of petals, a whorl of stamens, and a pistin (a whorl of carpels joined together).

Let us go back to the beginning. We believe today that flowering plants probably evolved in the Late Jurassic period, a little over 140 million years ago. We can say with certainty that many different forms of flowers existed in the Early Cretaceous (for details on fossil finds, see Friis et al. 2011). Only relatively few original flower forms have persisted over the course of flower evolution and are still with us today (Endress 2001). Among them are forms with numerous flower organs. The record holders are surely plants of the *Annona* genus (such as those in the custard apple family, *Annonaceae*, or the tropical fruit cherimoya, *Annona cherimola*). *Annona montana* has around 2000 stamens and *Annona muricata* approximately 500 carpels. This is slightly higher than the number of flowers borne by magnolias, tulip poplars, or plants in the *Schisandraceae* family.

If we now take a look at some very young flower primordia on a star anise plant (*Illicium anisatum*, Fig. 3), we notice that while the exact number of tepals (specialised for protection and visually attracting pollinators) and stamens may vary, the number of carpels is fixed at eight.

Taken together, however, all the organs form a consistent golden ratio spiral. In Fig. 3 top left, we see the formation of tepals in the bud, and bottom left the transition from tepals to the first stamens. On the right all organ primordia have formed on the growing tip (shoot apical meristem). An unusual feature here is the remnant of the central axis, which remains attached and grows thicker. It later holds together the fruit that we know as the star anise spice, and plays an important role in pollen tube competition.

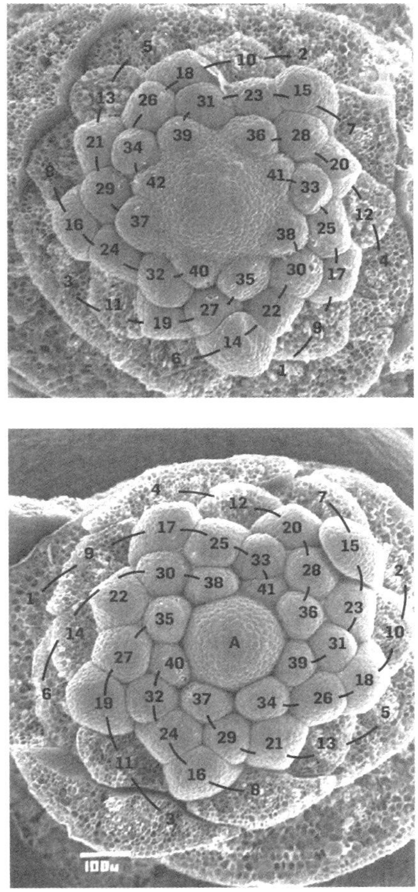

The larger the number of primordia arranged according to the limit divergence on the shoot apical meristem (or on the ›receptacle‹), the more clearly you can see oblique rows of primordia spiralling out from the centre. These spiralling lines are called parastichies. Arrangements with the parastichy number 8 (a number in the Fibonacci series) are particularly striking. In this arrangement, there are eight parastichies, each of which joins every eighth organ, moving from the base to the apex in the order of organ formation.

In many plants the golden ratio is applied very successfully in packing flower organs, yet it appears relatively rarely among flowering plants in general. How can we explain this discrepancy? The large number of pollen-producing stamens compensate for high losses from primitive pollen-eating insects – both those with chewing mouthparts, such as beetles and primitive butterflies (*Micropterigidae*), and those with licking mouthparts, such as flies (which do, however, pollinate the flowers ›in return‹). The original bait used to attract the insects was probably replaced relatively early on in the evolutionary development of flowers by much ›cheaper‹ nectar. At the same time, in a process of coevolution, the mouthparts of newly developing insects transformed into a sucking proboscis. Flowers became less complex (to save on pollen) and developed a whorled structure, frequently one in which the whorls alternate. Whorls formed as a result of ›rhythmisation‹, that is, the introduction of longer breaks in the chronological sequence of primordia. In many cases, a tendency to spiral can still be observed in individual whorls or all whorls.

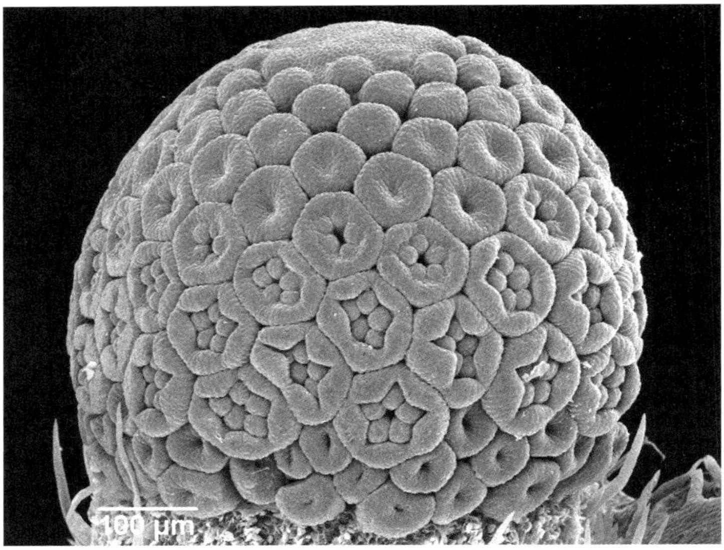

Figure 4. A view through a scanning electron microscope of a very young flower head of a daisy (*Bellis perennis*). The flowers at the very base (which were formed first) are initially held back in their development and later grow to become the white, female ray flowers which act as ›attention grabbers‹. All the flowers develop in a golden ratio sequence on the base of the head. As they grow older and larger, the five-tipped flowers come into contact with other flowers and, since every flower is surrounded by six others (of varying ages), they take on a six-sided shape.

It is particularly pronounced in the calyx (the collection of sepals). In roses, for example, it is easy to detect a spiral in the calyx of adult flowers. This was even the subject of a mnemonic verse attributed to Albertus Magnus (d. 1280): »Of five brethren born together, two had beards and two had none, and the third had but half a one«. The five sepals are, of course, not ›born‹ at the same time, but in a clear $^2/_5$ sequence with increasing simplification of the sepal edges.

Simplified flowers are often clustered together in inflorescences that in turn can have a spiral structure. For example, in almost all members of the *Compositae* or *Asteraceae* family – which, at 25,000 species, is the largest family of flowering plants – the golden ratio is again responsible for the very efficient way in which they pack flowers (or to be more precise, florets) in the flower head. When we look at a daisy (*Bellis perennis*), for instance, we can clearly see the spirals resulting from the golden ratio across the entire disc at the centre of the flower head. And if we examine the flower head in terms of its development over time, we discover another ›secret‹ of the golden ratio. Every floret on the head is surrounded throughout its development by six others of varying ages.

They soon start to press against each other and, as they grow larger, necessarily take on a six-sided shape (Fig. 4). There is no other shape more space-saving than a hexagon, as is similarly evident in the example which immediately springs to mind of the hives of colony-building hymenopterans (such as bees or wasps, see Fig. 1, p. 154).

Efficiency in dispensing pollen grains for the pollination of flowers can be closely connected with the golden ratio in the spiralling of multiple flowers in dense inflorescences or of multiple stamens in individual flowers. The way in which flowers or stamens release their pollen, both in terms of timing and relative spatial position, reflects the order in which they were formed, which in turn is determined by the limit divergence. Even simplified flowers, in which approximately five stamens are formed according to a $^2/_5$ divergence, can release their pollen grains in a corresponding sequence. Pollen portioning is almost universal among flowering plants, a fact which in itself indicates that portioning is more effective than a single discharge.

Beauty shaped by the golden ratio and the optimisation principle

It is perhaps precisely the connection between the golden ratio and the principle of optimisation that allows us, at an unconscious level, to experience a powerful aesthetic. In many ways, our thoughts and actions are guided by the dictates of efficiency, as they surely have been since our very beginnings.

Braun, Alexander: »Dr. Carl Schimper's Vorträge über die Möglichkeit eines wissenschaftlichen Verständnisses der Blattstellung, nebst Andeutung der hauptsächlichen Blattstellungsgesetze und insbesondere der neuentdeckten Gesetze der Aneinanderreihung von Cyclen verschiedener Maasse«, in: Flora 18 (1835), Regensburg, Königlich. bayer. botanische Gesellschaft.

Endress, Peter: »The flowers of extant basal angiosperms and inferences on ancestral flowers«, in: International Journal of Plant Sciences 162 (2001), Chicago, The University of Chicago Press.

Friis, Else M./Crane, Peter R./Pedersen Kaj R.: The Early Flowers and Angiosperm Evolution, Cambridge, Cambridge University Press 2011.

Hofmeister, Wilhelm: Allgemeine Morphologie der Gewächse, Leipzig, Engelmann 1868.

Leins, Peter/Erbar, Claudia: Blüte und Frucht. Morphologie, Entwicklungsgeschichte, Phylogenie, Funktion und Ökologie, Stuttgart, Schweizerbart'sche Verlagsbuchhandlung 2008.

Leins, Peter/Erbar, Claudia: Flower and Fruit. Morphology, Ontogeny, Phylogeny, Function and Ecology, Stuttgart, Schweizerbart Science Publishers 2010.

Schoute, Johannes C.: »Beiträge zur Blattstellungslehre«, in: Recueil des Travaux Botaniques Néerlandais 10 (1913), Nijmegen, Société botanique néerlandaise.

Golden Pythagoras trees
Fractals and self-similarity

Daniel Lordick

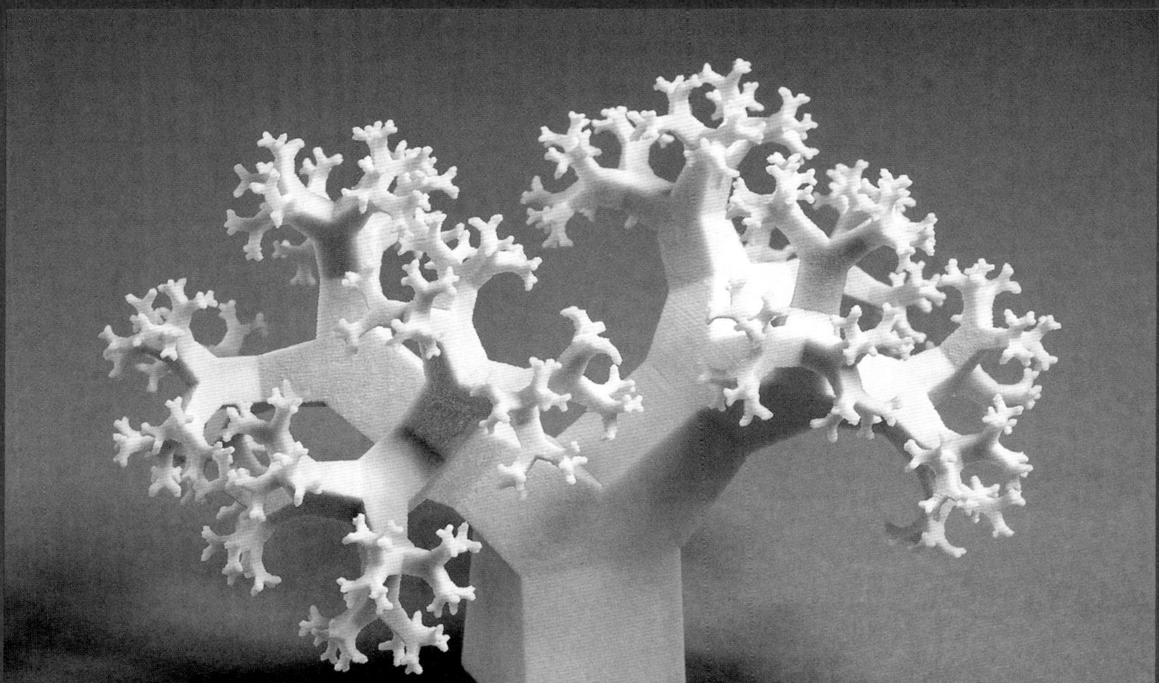

The universe, said Galileo Galilei, is written in the language of mathematics, and the mathematical concepts of fractals and the golden ratio seem to demonstrate the validity of that assertion for biology too. After all, fractals are perfectly suited to modelling such complex structures as fern leaves and blood vessels, and a remarkably precise approximation of the golden ratio can be found in the inflorescences of many plant species, where it appears as the golden angle (approximately 137.5°). This can be seen, for example, in the spiral arrangement of seeds on the head of a sunflower. The number of spirals is always a member of the Fibonacci series, and as the terms of the Fibonacci series become larger, the quotients formed by successive members approximate the golden ratio $\phi=(1+\sqrt{5})/2 \approx 1.618$.

In the following, we will experiment with combining the notions of *fractals* and *golden ratio*, which is not as outlandish as it may seem at first. After all, the tasty variety of broccoli known as ›Romanesco‹, available at any well-stocked vegetable stand, combines a fractal structure with spirals arranged according to the Fibonacci sequence (cf. Walser 2004).

Fractals are created in a sort of feedback process in which a pattern is repeatedly scaled down and copied (Peitgen 1992). Each application of this process of repeatedly scaling and copying is known as an ›iterative step‹. This method of construction results in a central feature of fractals: self-similarity. When we look at a fractal at various scales, in other words, the same picture appears again and again (scale invariance).

Figure 1. Y-fractal as three-dimensional analogue to the Pythagoras tree

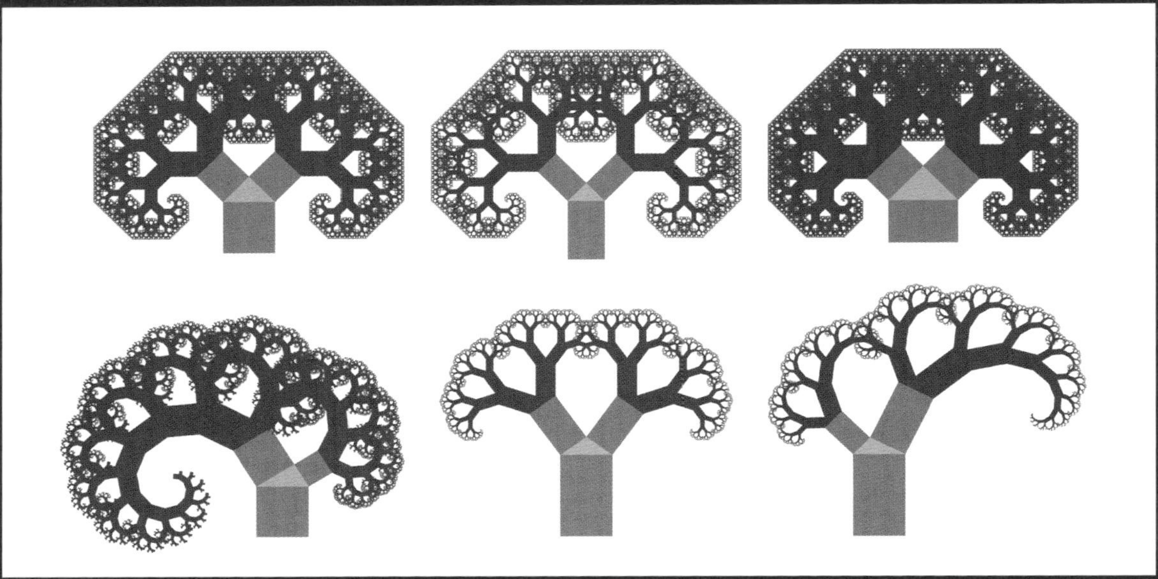

Figure 2. (top left) Pythagoras tree with isosceles triangles; (top middle and top right) Pythagoras tree with golden rectangles lying on their long edges or standing on their short edges; (bottom left) Pythagoras tree with triangle legs in the golden ratio; (bottom middle) symmetric tree fractal with scaling factor based on the golden ratio and a resulting branch angle of 72°; (bottom right) tree fractal with branch widths in the golden ratio and a branch angle of 72°. The figures are drawn to 14 iterations.

Self-similarity can also be observed in the golden spiral, in this case from segment to segment. This is not a fractal, however, since it can be created through a process of scaling down and copying no more than *once* per iteration with the factor $\phi-1 \approx 0.618$. The golden spiral thus necessarily lacks another typical feature of fractals, namely – as a rule – a fractional Hausdorff dimension (which through the Latin *fractus* gives us the term ›fractals‹).

Pythagoras tree and golden variations

Let us now take a popular plane fractal, the Pythagoras tree, and try to combine it with the golden ratio. By applying the golden ratio, we hope to create something new, something more pleasing than the initial figure. We will pursue this approach uncritically for the moment in order to explore the full range of possibilities. In addition to the two-dimensional figures, we will then create three-dimensional analogues, and finally we will construct a lush tree with the best possible connection with the golden ratio.

The starting point is the Pythagorean Theorem: a right triangle, initially an isosceles triangle, is set above a square, with the hypotenuse of the triangle forming the side of the square. In the first iterative step, we add two more squares over the legs. The sum of the areas of these squares equals the area of the first square, according to the Pythagorean Theorem. The scaling factor in this twofold process of scaling down and copying is $f = \sqrt{2}/2 \approx 0.707$. Repeating this process indefinitely leads to the Pythagoras tree (Fig. 2 top left). It follows from the way the tree was built that in each iterative step, the sum of the areas of the new squares is equal to the area of the initial square. Some of the branches start to overlap from the fifth iteration onward.

To ›gild‹ the tree, one trick – admittedly a cheap one – would be to replace the squares with golden rectangles, i.e. rectangles in which the sides are in the ratio $1{:}\phi$. That stretches or compresses the tree a little but does not change its basic shape (Fig. 2 top middle and right). For a fractal, after all, what is crucial is not the initial shape but the process of repeatedly scaling down and copying. Another variation involves using not an isosceles triangle but one in which the legs are in the golden ratio to one another (Fig. 2 bottom left). On the other hand, if we eliminate the requirement of the right angle and instead choose the scaling factor ϕ^{-1}, we obtain an isosceles triangle with an obtuse angle of 108°. This triangle forms part of a regular pentagon (Fig. 2 bottom middle). Here also, we can choose lengths for the shorter sides of the triangle that are in the golden ratio to one another. As in the case of the Pythagorean Theorem, which of course no longer holds in the present case, the point of contact of the two small squares can be thought of as traveling along the arc of a circle. Because of the Inscribed Angle Theorem, the obtuse angle of the triangle then remains a constant 108° (Fig. 2 bottom right).

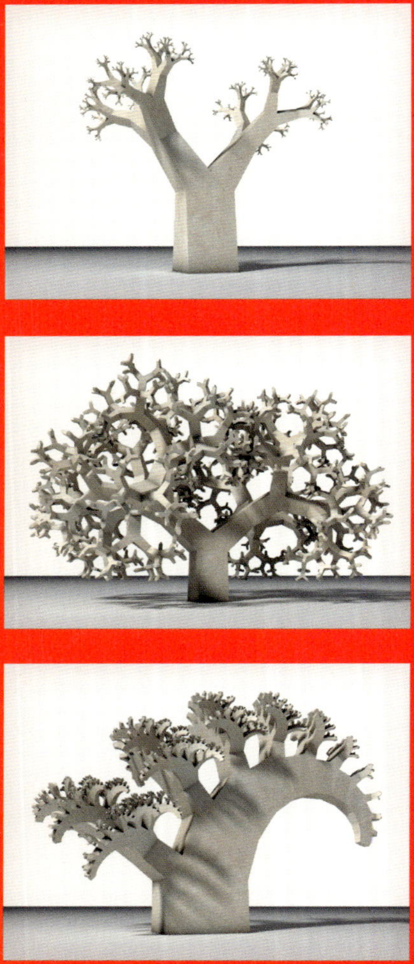

Figure 3. (top) Y-fractal with golden ratio as scaling factor and a branch angle of 72°; (middle) Y-fractal with surfaces scaled according to the golden ratio; (bottom) Y-fractal with golden rectangles as cross sections and a branch angle of 72°. The renderings show ten iterations. In all three fractals, the base objects used appear as golden rectangles from one perspective.

Three-dimensional interpretation of the Pythagoras tree

Let us now turn to the Pythagoras tree that we started with and interpret that tree in three dimensions. The natural thing to do would be to use cubes instead of squares and triangular prisms instead of triangles. Unfortunately, the edges of the upper square surface then lead to a rather angular appearance inconsistent with organic structures. We therefore apply a scaling factor to reduce the top surface of the initial cube to a rectangle in which the smaller side measures $\sqrt{2}/2$. Now, at each iterative step, the next generation is rotated by 90° to prevent self-intersections (Fig. 1, p. 164).

Three-dimensional variations using the golden ratio

As in the two-dimensional case, we use this tree as a starting point and carry out various modifications in order to elicit the effect of the golden ratio. But now that we have become accustomed to the lush crown, which may remind us of a lime tree, the result that we obtain when using the golden ratio as a scaling factor may be disappointing (Fig. 3 top). The boughs shrink faster than in an African dragon tree. One solution is to scale surfaces by the golden ratio instead of scaling edges by the factor ϕ-1 . The edge ratio of succeeding generations is then $\sqrt{\phi-1} \approx 0.786$, and with this scaling factor, the new tree is even more luxuriant than the Pythagoras tree. Although there are now self-intersections, this version may be the most appealing one incorporating the golden ratio (Fig. 3 middle).

Lastly, we will propose an asymmetric variant in which the base of the tree is a golden rectangle. Like the corresponding triangle in a regular pentagon, the polygonal base of a triangular prism has an obtuse angle of 108°. At each iterative step, only one branch is rotated by 90°. These specifications and the desire for a reasonably smooth tree lead to clearly defined reduction factors. This almost inevitably results in the version shown here (Fig. 3 bottom).

Concluding remarks

The golden ratio is only one of an infinite abundance of possible division ratios that can be used to vary the Pythagoras tree. Any other proportion could be used with equal justification. It follows that there are an infinite number of alternative configurations, some of which may be more pleasing aesthetically. No matter how surprising it is to discover frequent examples of the golden ratio in nature, exclusively using the golden ratio as a design tool clearly remains an arbitrary decision and one that is rarely successful.

Peitgen, Heinz-Otto/Jürgens, Hartmut/Saupe, Dietmar: Bausteine des Chaos. Fraktale, Berlin/Heidelberg/New York, Klett-Cotta/Springer 1992.
Walser, Hans: Der Goldene Schnitt, Leipzig, Edition am Gutenbergplatz Leipzig 2004.
For more examples of combining fractals and the golden ratio: http://blog.wolfram.com/2014/05/22/adventures-into-the-mathematical-forest-of-fractal-trees/, accessed: 25 May 2016.

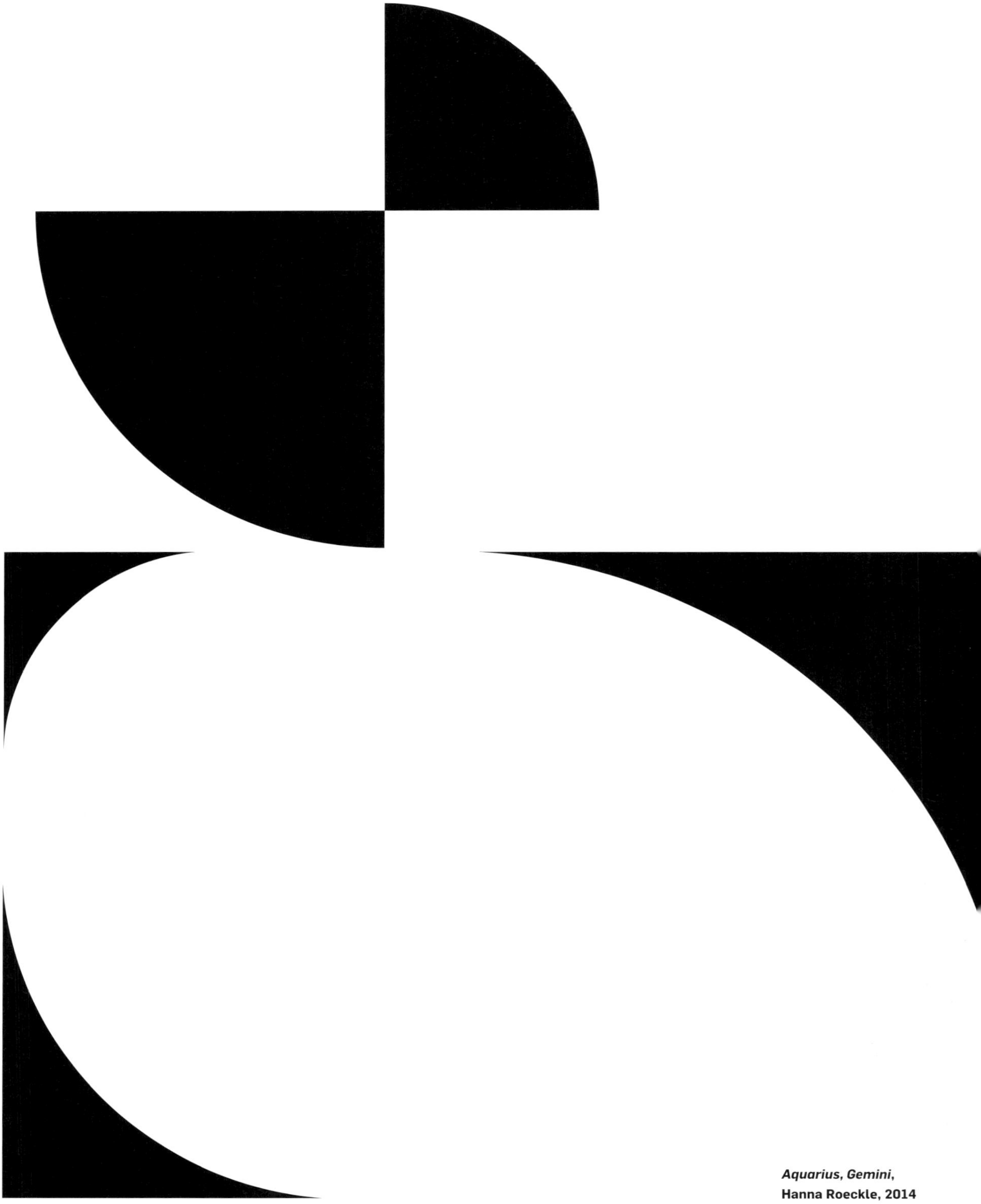

Aquarius, *Gemini*,
Hanna Roeckle, 2014

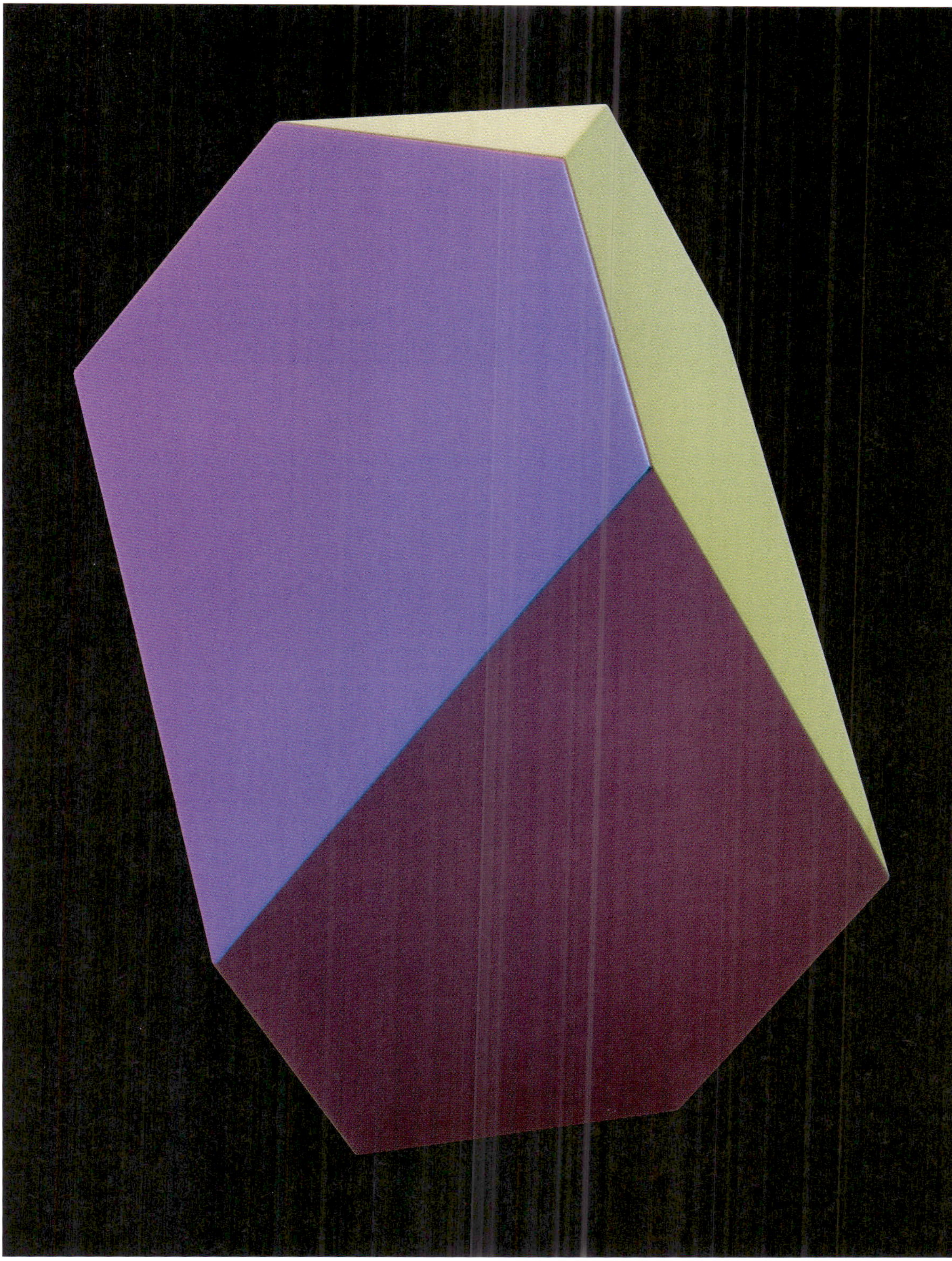

Proportions in music as a temporal art

An experiential or merely an abstract symbolic ascription?

Hartmut Fladt

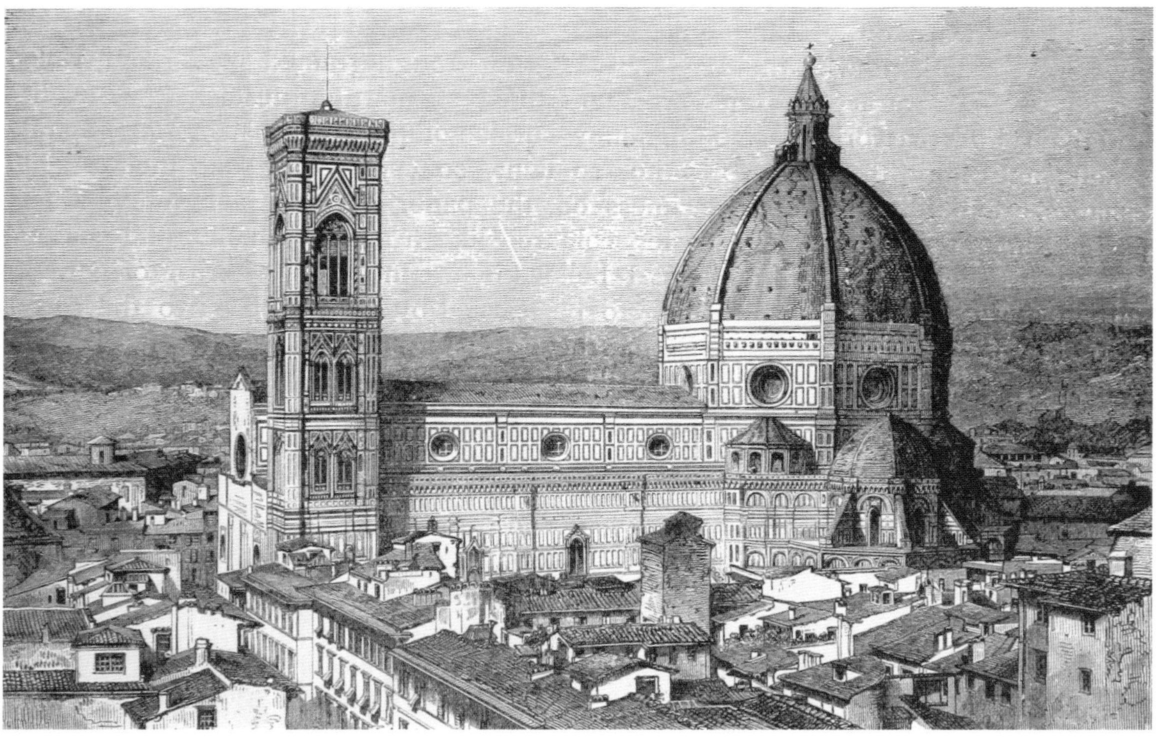

Figure 1. Florence Cathedral, Samuel Manning, 1870

» The Sun, in ancient guise, competing
With brother spheres in rival song …«

Even in the ›Prologue in Heaven‹ in Goethe's *Faust*, the fascination of a concept of music is conveyed which since antiquity – with Sumerian roots – has emphasised this art's mathematical side, its ›well-orderedness‹ according to numbers and proportions. This well-regulated state is primarily measured by way the tonal system relates. Musical intervals on string instruments, particularly the monochord, can be represented as whole-number proportions, produced by dividing the strings. This creates scales and sounds where consonances and dissonances simultaneously (as chords) and successively (as melodies, lines) form a *harmonia*, a unity of manifold diversities. Since this construct is also the basis of proportions in a great many other contexts, a concept of *musica* emerged which combined Pythagorean number theory with the dialectical theory of nature and went far beyond what we could today conventionally refer to as ›music‹.

In *Faust*, the harmony of the spheres of the *musica mundana* is described, the well-regulated order of the universe according to proportions defined by musical intervals (and hence for many years there was the theory that the planets – including the sun! – actually emitted ›tones‹ on their orbits). The *musica humana* then denotes the well-regulated order of the human body in its proportions. In the doctrine of the four temperaments, a well-ordered physical and mental state represented a *harmonia* in the mix of bodily humours, and this idea – which varied widely in its details – was similarly important for music and the emotions and passions it could represent, as well as for the ›music medicine‹ practiced from Late Antiquity to the Renaissance. Finally, *musica instrumentalis* corresponds to our own notion of music, encompassing everything involving the creation of sounds and tones with ›tools‹ and instruments (which also, naturally, include the voice).

Late Antiquity saw the development of the system of the seven *artes liberales*, not only providing a framework for thought, but also underlying the forms of organising culture up until and during the Late Renaissance. For our topic, the emergence of universities in Europe during the twelfth and thirteenth centuries plays a fundamental role. The *artes liberales* were taught as a *studium generale*, a basic course of study, across all faculties. As a result, those in the subjects which Faust says he ›alas‹ has studied, the prospective philosophers, lawyers, physicians and theologians, also studied music. As music mathematicians, Marchettus of Padua (c. 1320) and other composers and music theorists taught the Quadrivium (comprising music, astronomy, arithmetic and geometry) at cathedral schools and universities during the Late Middle Ages and the Renaissance. Johannes Ciconia (Padua/Venice c. 1400), who was one of the first Franco-Flemish composers in Italy, founded a musical tradition lasting 200 years, wrote mathematical-musical treatises such as *De proportionibus*. This ›ennoblement‹ of music presented a challenge to representatives of the ›mechanical‹ or applied *artes*, which were seen primarily as craft skills. Artists, sculptors and architects (often combined in one person, such as Brunelleschi in Florence and later Michelangelo in Rome) also strove to integrate the ›noble‹ proportions and symbols into their works. When artists created subjects such as Raphael's *School of Athens* fresco, their works were not merely intended to attest to their erudition, but also evidence, just as in their experiments with perspective, how they had mastered the symbols and the weighty principles of proportional theory.

In the Trivium, which comprised the three language-related *artes* (grammar, rhetoric and logic), music had a special status, since it belonged both to this lower division of the seven liberal arts as well as to the Quadrivium, the higher division. In the latter, music was a mathematically based art of proportions.

On the other hand, as sung language and an artificial musical expression of words, it always references its shared roots with language itself. Every choral work, for example, has textual punctuation congruent with the musical punctuation – the punctuation marks also correspond to musical cadences. There was a fluid transition to the musical rhetoric of antiquity (particularly Quintilian, whose system was still unquestioningly taught at the Latin schools and grammar schools in Germany in the 1800s). Even well into the nineteenth century, purely instrumental music could be structured by formal rhetorical principles.

In which levels of music are proportions concretely realised?

1. The tonal system is fundamental to all musical and supra-musical constellations, whether the basis of calculation is the Pythagorean tetractys (1:2:3:4) or the geometrical relation of the golden ratio a:b = b:(a+b) and the Fibonacci sequence (1, 1, 2, 3, 5, 8, 13, 21, 34, 55, 89, 144 etc.). However, in both cases there were and are problems reconciling them with the physical principles of the partial tone series, which are exactly subsumed in the whole-number proportions of the intervals of string division (1:2:3:4:5:6:7:8:9:10 etc.). The irrational number 0.618... as an approximate factor of the ›golden‹ division of lines roughly corresponds to the ratio of 5:8 (the minor sixth interval), which was interpreted as a bridge between the geometric relation and the whole-number proportions of the Fibonacci sequence. As an arbitrary ›Fibonacci selection‹ from the sequence of partial tones, at least 2:3:5:8 produces a meaningful tonal result (C^1 from C-G-e-c^1), but would be unable to allow for other substantive elements. The tetractys as a ›model of the world‹ is limited by the need to derive everything solely from the numbers from 1 to 4, and so in music it is compelled to derive all other intervals from the octave (1:2), the fifth (2:3) and the fourth (3:4).

Consequently, for example, the major third (as a pure 4:5 interval) has to be calculated as a doubling of the whole tone, which in turn is defined as the difference between the fifth and fourth – the result is a decidedly unmelodic interval too large by a ›syntonic comma‹. The minor third is correspondingly too small, the half-tone as the difference between the fourth and major third is too small, and so on. The Pythagorean ›heptatonic diatonic‹ system can also be represented as an adjustment of pure fifths. Ordered as a scale, the seven tones F^1 – C – G – d – a – e^1 – b^1 represent the ›white keys‹ on the piano. Since the pure minor triad 4:5:6 is not possible, sixteenth-century theorists and composers developed a ›scenario‹ based on the numbers 1 to 6. Pythagorean tuning was replaced by musical tuning systems such as the ›mean tone‹ (from around 1450 to approximately 1750) and differently tempered tunings (from around 1700 to the present).

In addition, all subsequent attempts to harmonise the whole-number proportions of the partial-tone series with the tetractys and the Fibonacci sequence were and remain wishful thinking. In particular, in the history of ›music and numbers‹ and the ›harmonic view of the world‹, it constantly needs to be borne in mind that rigorous mathematical derivation and immanent logic do not automatically exclude ideas and ideologies rooted in magic. Both the metaphysical premises and the corresponding conclusions of these logical immanences per se do not have to follow a particular external logic. How can Mephistopheles – conceived by Goethe as Faust's necessary ›particular negation‹, his ›own other‹ – be exorcised as a representative of ›evil‹? As Faust asks, or perhaps more accurately, states, »The Pentagram gives you pain?« (Part 1, The Study). The pentagram, or ›druid's foot‹, is regarded as the most perfect geometric figure since it consists exclusively of golden ratio proportions. But, given that this ›beauty‹ is completely abstract, why should evil not be enamoured of it? After all, concentration camp commanders demonstrably loved Mozart and Schubert and were ready to deny the moral character of this music. Blood-stained dictators entertained and continue to entertain lofty notions of ›purity‹, including ›racial purity‹.

2. The concept of music as a temporal art came to play an increasingly dominant role in the history of discussing the relevance of the Fibonacci proportions. Rhythms, measures (time signatures), metres of all kinds, tempo relationships, syntax and form as time structure were then able to challenge musical thinking when the further development of ›mensural notation‹ (14th/15th century) facilitated the precise notation of the smallest musical time units, whether as a tone or a pause, as well as the most complex proportional relationships. However, the factor 0.618… was and is applied in calculating the main dividing points of musical forms, i.e. in organising ›filled time‹ (from motivic to thematic, harmonic, gestural, and dramaturgical). This is repeatedly evident particularly with smaller two and three-part types of form from many eras across musical history. Even forms primarily conceived in rhetorical terms can additionally represent proportions of the golden ratio. However, this should not be seen as implying a general ›law of nature‹ – as long as it does not become an ideology, it remains one meaningful possibility among many.

3. Finally, magical, ritual and symbolic aspects should not be forgotten. Ultimately, everything can be counted. What is counted (notes, pauses, beats, words, syllables, letters and so on – alone and in relation to one another) is often then extrapolated to the number alphabets of diverse gematrias (i.e. the interpretation of words with the aid of numbers). This opens the door wide to symbolic, mostly religiously motivated ›secret numbers‹ or to magical names and, indeed, invocations. There is no objection to the use of such approaches to document verifiable historico-philological relations and ways of thinking; however, there is a high risk of interpretative arbitrariness and obscurantism. In this regard, Ruth Tatlow's has exemplarily separated scholarly-based ›wheat‹ from a great deal of ›chaff‹ in the case of Johann Sebastian Bach.[1]

›Experienced time‹ and ›experiential time‹

Music operates inexorably over time. Most of its experiential qualities are linked with the phenomenon of ›time‹, whether via rhythm, tempo and form, or the dramaturgy of musical figures, gestures and the development of feelings. Music moves forwards. Yet then we experience symmetries, correspondences, proportioned equivalences, repetitions. Forwards yet back? In Fibonacci proportions? Or put another way, in the midst of a musical processuality, a moment of ›re-minding‹ occurs, a historical dimension of depth, an allusion to something long past which has nevertheless remained alive in the memory – of the composer, the interpreter, the audience. Forwards yet back? We encounter the phenomenon of the ›simultaneity of the non-simultaneous‹ in every piece of music, but also of course in the consciousness of the interpreter and recipient. Rather than remaining abstract, all this is vividly experienced and connected with a spectrum of individual associations.

Here, we are dealing here with ›experienced time‹ and the ›time of experience‹, *temps durée* and *temps espace* (using the categories developed by Belgian philosopher Henri Bergson), with the organic-processual and the spatial-architectural in the musical concept of time, in which the proportions of the golden ratio and Fibonacci numbers always play a role. We are dealing with what Igor Stravinsky described as ›psychological‹ and ›ontological‹ time.[2] The musicologist Carl Dahlhaus wrote: »The moments that Bergson bluntly separated from one another are not isolated but linked in experience by reciprocity. If *temps espace*, the empty before and after, is an abstraction of *temps durée*, then, on the other hand, the prolongations and shortenings of experienced time are only comprehensible against the background of the spatial. And both moments, *temps espace* and *temps durée* are, as time structure and movement, operative in music.«[3] For the temporal art of music, there is a substantial difference between objectifiable, spatialised time and experienced time. The musical experience may register particularly eventful, enthralling or deeply moving parts of a piece of music which, for example, only last 60 seconds as equivalent to another, uneventful section that is objectively five times as long. All this places a question mark over the epistemological weight of numerous abstractly ›coherent‹ Fibonacci relations.

Guillaume Dufay's *Nuper Rosarum Flores*, an ›isorhythmic‹ motet composed in 1436 for the consecration of Florence Cathedral, attained particular prominence precisely because of its connection between architecture and music and its multifaceted symbolic integration of the two spheres. Such an important international event as the cathedral's consecration required music featuring a high degree of spiritually symbolic, mathematical and gematric relationships and references. The 6:4:2:3 tetractys proportions of Solomon's Temple mentioned in the Bible were integrated into the cathedral architecture (Fig. 1, p. 172) and are reflected in Dufay's four-part work in the proportions of the four measures – and thus can be heard and experienced. The Fibonacci dimensions of the famous dome (144, 89, 55) can be counted out in the duration of the tonal units for the two defining tenor voices.[4]

›Fibonacci‹ as a guide through the obscure zones of cultural history

We have now reached the boundaries to the speculative, yet how and what is ›counted‹ and to which gematric systems symbols or magical invocations are ascribed remains – as a journey through the obscure zones of cultural history – nevertheless fascinating.

Béla Bartók was an admirer and collector of Fibonacci-structured products of nature, and this predilection is reflected in a number of his works. With its striking analysis of the role played by Fibonacci relations in the formal divisions of the fugue as well as in the instrumentation and tonal colours, Ernö Lendvai's discussion of the first movement of Bartok's *Music for Strings, Percussion and Celesta* (1936) is quite remarkable despite its shortcomings (for example, ignoring the very different durations of time signatures or Bartok simply adding an empty bar to reach the number 89).[5] The coda, however, with its magical and ethereal tonal web produced by celesta, tremolos and the original thematic figure played in the original register at the same time as its inversion in the highest register resolutely refuses to be located, even approximately, in the system of the golden ratio. Undoubtedly, Bartok ›positioned‹ Fibonacci orientation points, in their abstract numerosity, as bar numbers. However, in terms of aesthetic perception and compositional analysis they are not primarily experienced as proportions, but as the start, culmination and turning points of dynamic musical processes which, moreover, appear even more vivid because they are structurally based and marked by changes in tonal colour. In this context, ›Fibonacci‹ provides perceptible – and attractive – additional information.

1 Ruth Tatlow, Bach and the Riddle of the Number Alphabet, Cambridge 1991, pp. 1–36.
2 Igor Stravinsky, Poetics of Music in the Form of Six Lessons, Harvard 1970.
3 Carl Dahlhaus, Musikästhetik, Cologne 1967, see the discussion of concepts of time in Bergson, Husserl and Hartmann in the chapter »Zur Phänomenologie der Musik«, pp. 111 ff.
4 Hans Ryschawy and Rolf W. Stoll, »Die Bedeutung der Zahl in Dufays Kompositionsart: ›Nuper rosarum flores‹«, in: Heinz-Klaus Metzger and Rainer Riehn (eds), Guillaume Dufay (= Musik-Konzepte 60, edition text+kritik), Munich 1988, pp 47 ff.
5 Ernö Lendvai, »Einführung in die Formen- und Harmoniewelt Bartóks«, in: Bence Szabolcsi (ed.), Béla Bartók. Weg und Werk, Schriften und Briefe, Kassel/Basel/Munich 1972, pp. 117 ff.

The European and American avant-garde movements during the 1950s and 1960s (serial and post-serial phases) represent a highpoint in the use of Fibonacci proportions. In 1953–54, Iannis Xenakis, who was influenced by Le Corbusier and also published essays on music and architecture, composed his *Metastaseis*. Luigi Nono used Fibonacci relations, for example, in *Il canto sospeso* (1956), his statement of his political convictions. And finally Karlheinz Stockhausen investigated the proportions of the golden ratio in compositional and analytical-theoretical terms in numerous works exhibiting markedly esoteric tendencies. However, having now become a ›teachable‹ subject matter in the fields of composition and theory, these approaches, as methods among many others, have lost all traces of their esoteric character.

Dahlhaus, Carl: Musikästhetik, Cologne, Hans Gerig 1967.
Lendvai, Ernö: »Einführung in die Formen- und Harmoniewelt Bartóks«, in: Bence Szabolcsi (ed.), Béla Bartók. Weg und Werk, Schriften und Briefe, Kassel, Bärenreiter/DTV 1972.
Möller, Hartmut: »Goldener Schnitt«, in: Ludwig Finscher (ed.), Die Musik in Geschichte und Gegenwart, Sachteil 3, Kassel, Bärenreiter/Metzler 1995.
Neuwirth, Gösta: »Erzählung von Zahlen«, in: Heinz-Klaus Metzger/Rainer Riehn (eds), Josquin des Prés (Musik-Konzepte 26/27, edition text+kritik), Munich 1982.
Ryschawy, Hans/Stoll, Rolf W: »Die Bedeutung der Zahl in Dufays Kompositionsart: ›Nuper rosarum flores‹«, in: Heinz-Klaus Metzger/Rainer Riehn (eds), Guillaume Dufay (Musik-Konzepte 60, edition text+kritik), Munich 1988.
Steinhauser, Ulrike: »Musik und Architektur«, in: Ludwig Finscher (ed.), Die Musik in Geschichte und Gegenwart, Sachteil 6, Kassel, Bärenreiter/Metzler 1997.
Stravinsky, Igor: Poetics of Music in the Form of Six Lessons, Harvard 1970.
Tatlow, Ruth: Bach and the Riddle of the Number Alphabet, Cambridge, Cambridge University Press 1991.

Viewing in flux
Screen sizes

Florian Schütz

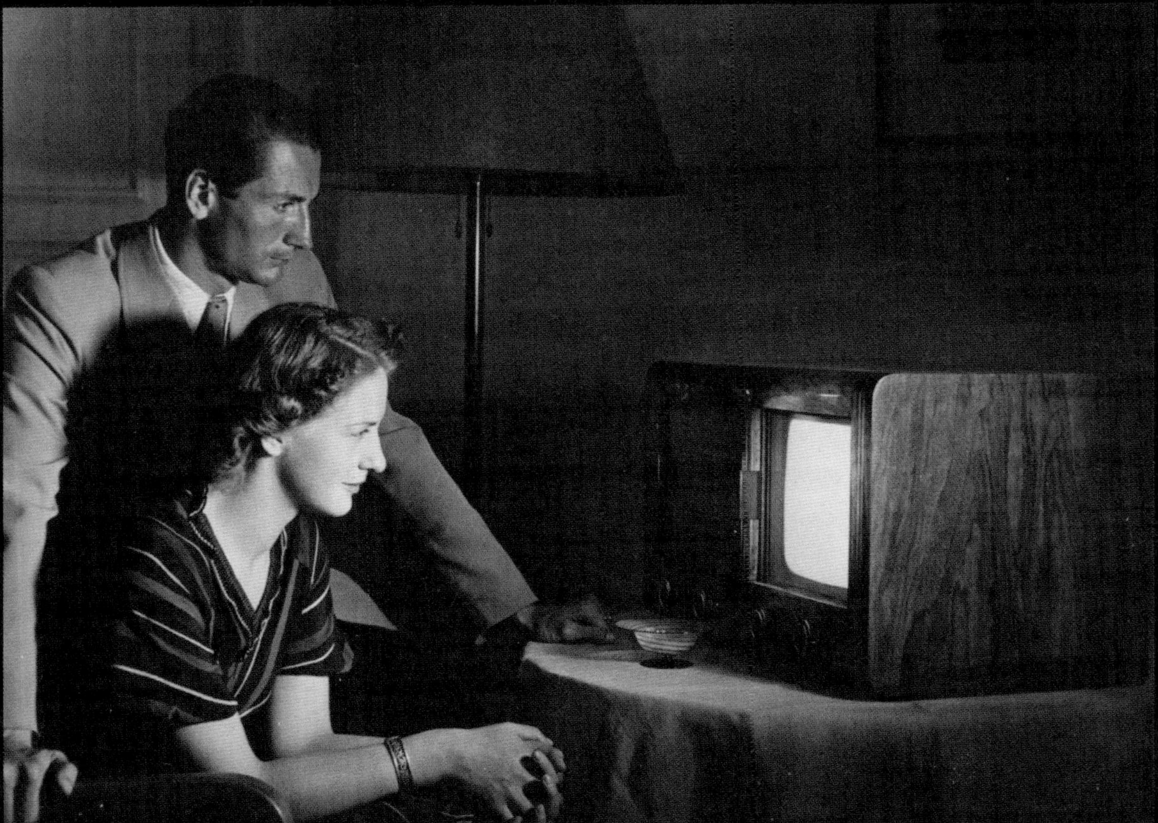

At the start of the millennium, over 90 % of households in Europe owned at least one television set. Initially, the first sets were so expensive that, in some cases, it was a question of either investing in a new small car or a TV. However, since its commercial success in the mid-1960s and subsequent mass distribution, television has significantly influenced daily and domestic life. Over those years, the TV has come to play a special role in people's lives, with ›home sweet home‹ being totally designed around the TV screen. The living room furniture is arranged around the TV, and the family meets at predetermined times in front of the screen to watch eagerly anticipated programmes. The television has become something between »cheap entertainment, culture and self-improvement«.

Yet rather than the history of moving pictures starting in the living room, its origins lie in and on different formats. The first film screenings were fairground attractions on offer in the late nineteenth century. Later on, films or newsreels were shown on cinema screens. Moreover, as the forms presenting these moving images have developed, from stamp-size experiments in the pioneering days to spectacular 3D shows in the multiplex mega-complexes of the twenty-first century, the consumers' viewing habits have also changed. Not only does the sheer size of the projection surface evoke a particular and unique visual experience, but the screen's aspect ratio also determines the area of film that is visible.

The first silent black-and-white films on mechanical television were shown on 4 x 4 cm-displays, with an aspect ratio of 1:1, and required a magnifying glass to even see the image at all. When the first mass-produced devices were issued, the television industry speedily agreed on the standard 4:3 screen, derived from the early cinema format. This movie format, though, was itself based on the universal standard of the film's physical size of 35 mm wide. As a result, the aspect ratio of the projected image in cinemas was 1.33:1 (approximately 4:3). Yet while the flexible cinema screen sizes rapidly produced changes in movie formats so that seating could be optimised in terms of distance to the screen and on the sides of auditoriums, the TV screen at home statically enclosed in wood or plastic remained bound to a 4:3 aspect ratio for many years.

To show films with widescreen projection formats on the almost square 4:3 screen, the visible area needs to be aligned, leaving a black strip at the top and base of the screen. Since the resulting rectangular format is rather like watching the film through the slot in a door used for postal deliveries, this practice of transferring widescreen to the 4:3 format is often called letterboxing. The analogous term pillarboxing is used to describe 4:3 content displayed on wider screens such as, for example, the 16:9 format. In this case, the lateral black bars on the right and left of the image create a shape reminiscent of the pillar post boxes originally used in the UK. If the widescreen images are not scaled to fit the 4:3 screen, the visual details on the cropped edges are lost. As a result, actors during filming are commonly placed more to the centre of scenes so they are not left out of shot when the movie is transferred to other formats. The 16:9 aspect ratio, though, is not derived from some potential maximum width of cinema screens. Instead, it is far more a compromise solution, since it neither loses too many details when transferred to 4:3 nor requires extensive adaption. Many modern cinemas show widescreen movies with a 2.35:1 format, which would correspond to a 21:9 aspect ratio on TV, leaving a highly compressed letterbox image on 4:3 TV sets.

For years, though, the television industry has tirelessly strived to create the cinema experience at home. Large cathode ray tube TV sets, not only very heavy but bulky, are now outdated thanks to increasingly flat plasma, LCD or LED screens. Display size, i.e. the length of the screen's diagonal, has been constantly growing, and the potential is certainly there for further significant increases. At present, formats such as 16:9, 16:10 or even 21:9 are far from rare, in both the TV and computer screen sectors. The ultra-thin curved screens with high pixel density are now fuelling a movement to a completely new viewing experience.

With the rapid ascendancy of smartphones and the constant multimedia access they provide, another key player has entered the ring in the battle over formats. Never has it been easier to take short videos anywhere and at any time, and circulate them within seconds. However, since most smartphone users hold the device in an upright position, the videos are recorded in the vertical portrait format. Although the most common aspect ratios of smartphone displays – e.g. 3:4, 3:5, 9:16 – would allow horizontal smartphone videos to be shown adequately on standard screens, ›vertical videos‹ are omnipresent in cyber culture and have generated much heated debate and fierce condemnation. In viral videos ranging from tutorials to rants, internet users have vented their feelings on how vertical videos upset their sense of aesthetics. There are good reasons, they argue, why TVs have widescreen formats, and vertically filmed content is totally at odds with that format. Smartphone producers have reacted to this trend and now ›warn‹ users when the video function is activated with the camera in a vertical position, advising them to switch to the landscape format. But the question remains – will our visual habits change fundamentally if such pedagogical measures are ignored?

Coincidence in progress
Leipzig's Old Town Hall

Marko Kuhn

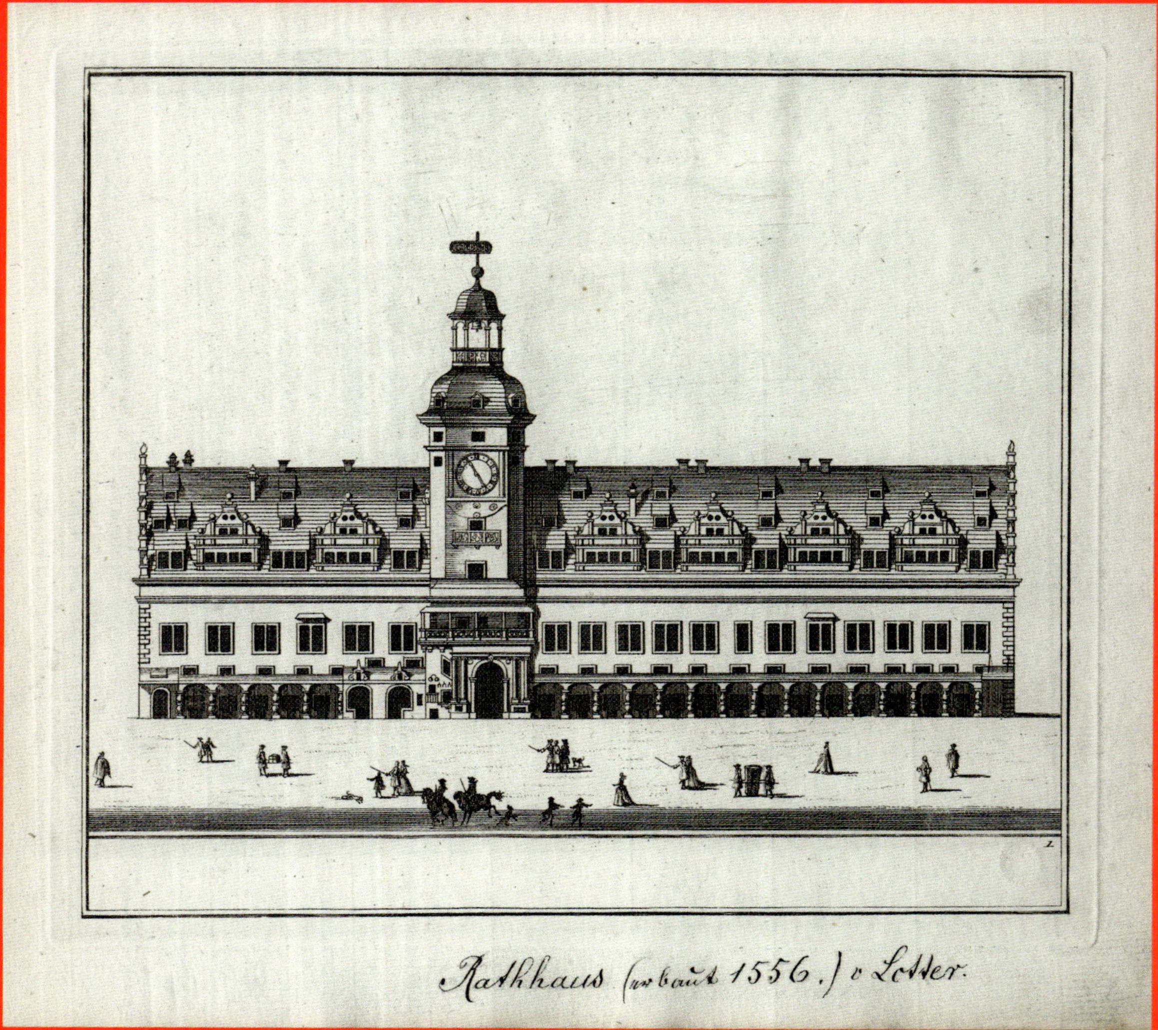

Leipzig's Old Town Hall (*Altes Rathaus*) is rightly ranked among central Germany's most beautiful and impressive buildings. Regarded as one of the major architectural works of the German Renaissance, it is cited in numerous scholarly works as a prime example of the golden ratio. With a total length of over 93 metres, the Old Town Hall is imposing in itself and, thanks to the no less impressive size of the Market square in front, it can be seen in its entirety, a view which further underscores the building's ›perfect proportions‹.

Just sheer coincidence

The Old Town Hall is indeed an ideal example of this subject, although not for the reasons so often assumed. First of all, the building's offset tower does not mark the golden ratio. Instead, the line of the golden ratio actually runs through the centre of the main portal, set asymmetrically in relation to the tower (Fig. 1). Second, the golden ratio in the architecture of the Old Town Hall was, in fact, sheer coincidence. When the Town Hall was refurbished in 1556, the conversion work was more or less only to revise the appearance of a complex of various individual buildings which had become fused over the years to create a unit. Certainly, no one at that time intended – consciously or unconsciously – to construct the Old Town Hall around the golden ratio.

The history of the building down the centuries

The written records first mention a town hall in Leipzig in 1360, although neither its size nor location is specified. Scholars assume, though, that even then it stood on the eastern side of today's Market square (in those days, still a clay pit). By 1341 a Romanesque-style house had already been built there, since in that year Friedrich II, Margrave of Meissen, gifted the building to the Leipzig cloth-makers. This prestigious two-storey building of massive stone had probably already been completed by the late thirteen century.

Figure 1. The Old Town Hall (*Altes Rathaus*), Leipzig, c. 1722

In this era, the cloth trade and cloth-making accounted for the largest sector in the town's economy, and the eminent members of the cloth merchants' guild held the highest civic offices. Most likely, then, the guild house was also used for Leipzig's city council meetings. The cloth hall's floor area was approximately the same as the council meeting room in the Old Town Hall today. The only surviving element above ground from the Romanesque guild hall is a window discovered in 1906 and set into the south side of the Old Town Hall.

As Leipzig grew in size, so did the tasks which fell to the city council. The cloth hall was enlarged a number of times to the north, building on foundations from previous structures. This, for example, also explains the slight ›kink‹ in the longitudinal façade of today's Old Town Hall. The surviving plans of the cellar vaults and remains of the foundations suggest that these extensions comprised two buildings stretching to today's pedestrian passageway through the Old Town Hall. In the mid-fifteenth century, these structures may well have been connected to two buildings on the other side of the passageway between the Market square and the small Naschmarkt square at the rear of the building, turning the Town Hall into a unified complex. At some point, a stair tower was constructed over the passageway. Although we do not know exactly when the tower was built, it must have been some time before 1476, since it first appears in written records at that time – and this, then, would have already set the proportions of the golden ratio for any later refurbishment.

In 1467, the council chamber was created with its present dimensions, as was the council hall later. In this process, the ceiling heights were aligned, since they varied in the individual partial buildings. In 1483, a secret chamber reached via a narrow staircase was constructed in a kind of mezzanine floor. This was set below the council chamber, and an opening in the floor allowed valuables and important files to be dropped directly into it. The refurbishments to the Town Hall at this time also included, not least, a façade with a unified design. Some years later, the cloth makers and merchants had a new guild hall constructed at a different location and moved there.

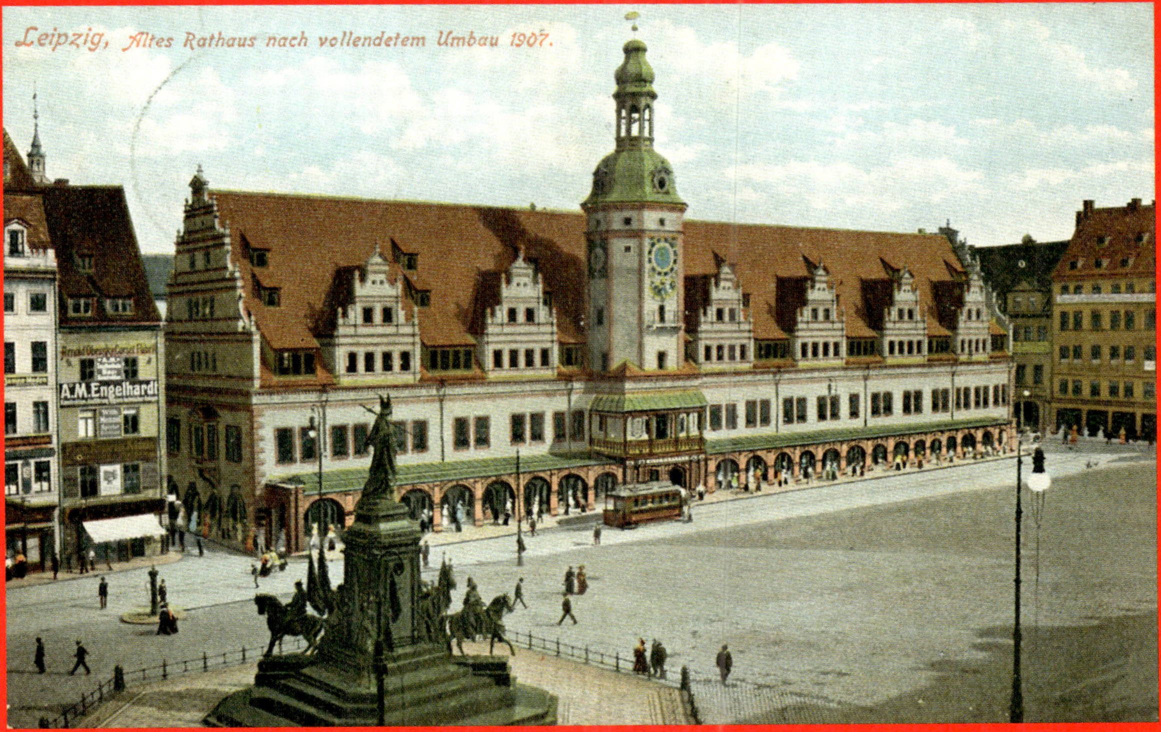

Figure 2. Postcard of the refurbished Town Hall, after 1907

By the mid-sixteenth century, the Old Town Hall had undergone numerous other refurbishments and conversions; the largest of these measures involved widening the building by around four metres towards the Naschmarkt square and, at the same time, enlarging the roof and increasing its height.

In 1556, under Hieronymus Lotter, the Old Town Hall was largely given the shape it has today. However, in terms of the golden ratio, this work merely converted and refurbished the medieval town hall. A new scaffold stair tower with an octagonal upper section was constructed on the foundations of the old tower, the building was given a total of thirteen large gables, the design of the windows unified, and ornamental architectural elements added – and, in this context, the portal facing the Market square was a particular highlight. Lotter was then mayor of Leipzig and responsible for the project, but is still often wrongly described as the architect of the Old Town Hall. This legend was fuelled, not least, by Lotter himself. In 1573, he left in the tower a document, written in his own hand, presenting himself as the building's sole architect. According to the latest research, however, the real architects were Paul Speck and, after his death, Paul Widemann.

Just as in earlier times, the Old Town Hall was not solely the seat of the municipal administration, but also the location of the courts. In 1587, for example, it housed the supreme court, the court of aldermen and the city court. In addition, it was the site of the council archives, and part of the cellar was used as prison cells. The large council hall, over 40 metres long and the largest hall in the city in its day, was used for all kinds of festivities. Traders and merchants continued to store goods in the underground vaults, and later also had stores and sale rooms on the ground floor.

Over the subsequent decades, the work on the Old Town Hall was primarily limited to enhancing the interior furnishings and fittings. In the course of such measures, for instance, the series of ruler portraits was updated. The original works, hung in the council chamber and banqueting hall, had been completed by Hans Krell in 1553. The exterior was thoroughly renovated in 1672, with an inscription added that circles the entire building.

The inscription, which is still visible today, measures 220 metres in total. In 1744, the Town Hall tower was then extended to its present height of 41 metres.

The Old Town Hall becomes a historical location

By the nineteenth century at the latest, the Town Hall was no longer able to cope with the demands of the rapidly growing city. Furthermore, in structural terms, the building was in a deplorable state. In the early 1880s, a strong and resolute initiative called for the demolition of the Old Town Hall (and the Old Stock Market as well) and the construction of a new town hall on the same site. With the city council deterred from such a policy, first and foremost, by the costs, the decision was taken instead to construct a new town hall at a different location. In this era, there was also a growing appreciation of this historic building's importance. Between 1906 and 1909, after the municipal administration had vacated the building, the Old Town Hall was thoroughly refurbished and renovated (Fig. 2, p. 189). However, there were still no definite plans for its future use. Finally, in 1908, the council decided to house the local history museum (*Stadtgeschichtliches Museum*) in the Old Town Hall. The museum moved in when renovation work was completed in 1909 and has been located there ever since.

Between 1943 and 1944, Allied bombing raids completely destroyed the roof's timber framework and a large part of the tower. It was not until 1950 before reconstruction work started. Between 1988 and 1990, the *Stadtgeschichtliches Museum* was closed for conversion and refurbishment. At present, two floors of the Old Town Hall showcase the permanent exhibition, which is regularly updated, recounting the history of Leipzig to the present day. The ceremonial hall is still an impressive setting for important events. Moreover, given the Old Town Hall's historical importance, this remains the venue where official guests enter their names in the guest book of the city of Leipzig.

Albrecht, Stephan: Mittelalterliche Rathäuser in Deutschland. Architektur und Funktion, Darmstadt, Wissenschaftliche Buchgesellschaft 2004, pp. 169–171.
Günther, Wolfram: »Hieronymus Lotter«, in: Arnold Bartetzky (ed.), Die Baumeister der »Deutschen Renaissance«. Ein Mythos der Kunstgeschichte?, Beucha, Sax-Verlag 2004, pp. 73–110.
Mundus, Doris: Das Alte Rathaus in Leipzig, Leipzig, Lehmstedt 2003, pp. 8–14, 31–50.
Posern-Klett, Karl-Friedrich von (ed.): Urkundenbuch der Stadt Leipzig, Vol. 1 (Codex diplomaticus Saxoniae Regiae II/8.1), Leipzig, Giesecke & Devrient 1868, pp. 24-25.
Rodekamp, Volker (ed.): Das Alte Rathaus zu Leipzig, Altenburg, DZA 2004.
Schwarz, Alberto: »Das Alte Rathaus vor der Erneuerung durch Hieronymus Lotter«, in: Marcus Cottin/Doris Mundus (eds.): 450 Jahre Altes Rathaus zu Leipzig. Neue Forschungsergebnisse, Beucha, Sax-Verlag 2009, pp. 23–40.
Walser, Hans: Der Goldene Schnitt, Leipzig, Edition am Gutenbergplatz 2004, pp. 133–134.

Pendant light *Golden Ratio*,
Lutz Koebele-Lipp, 2016

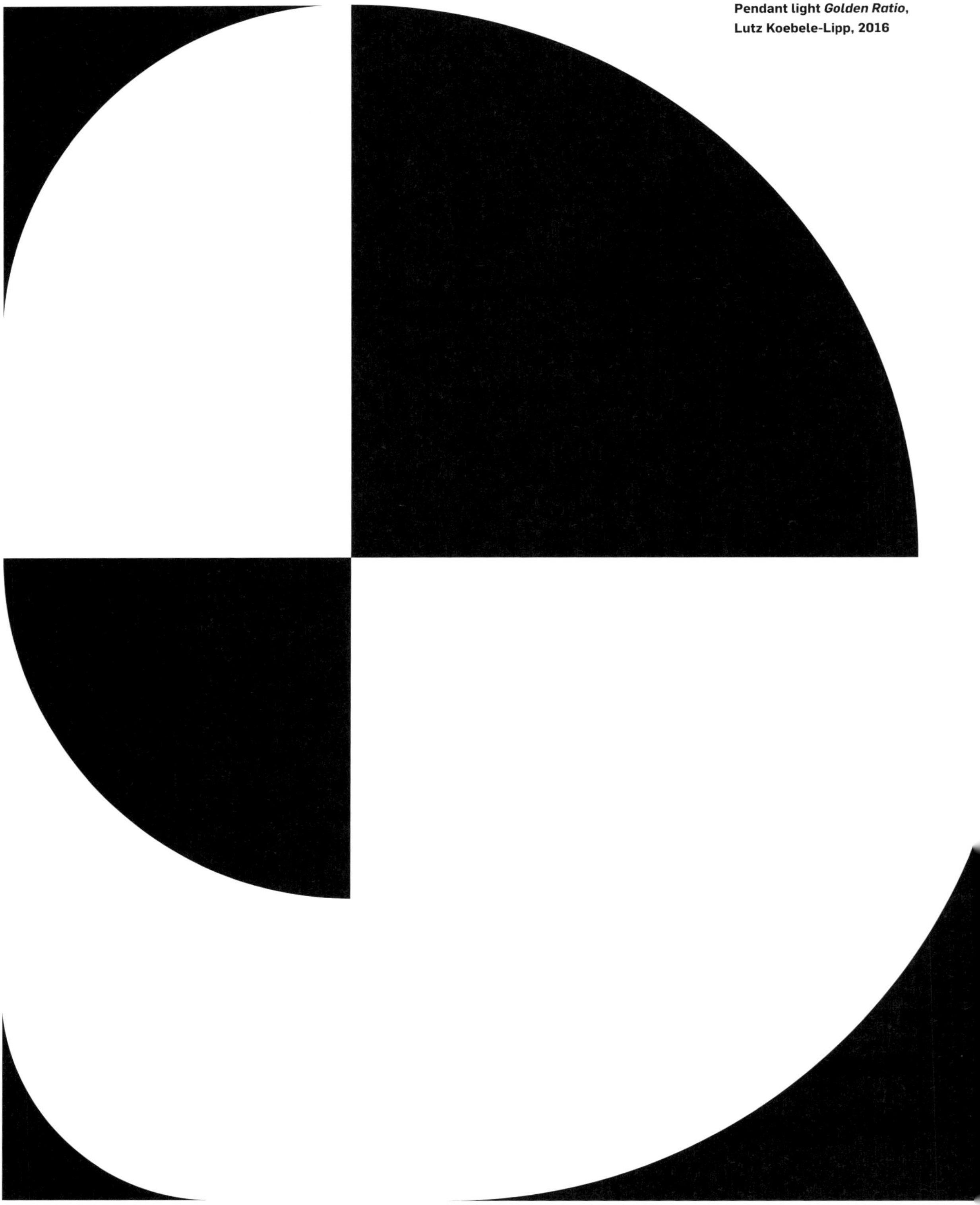

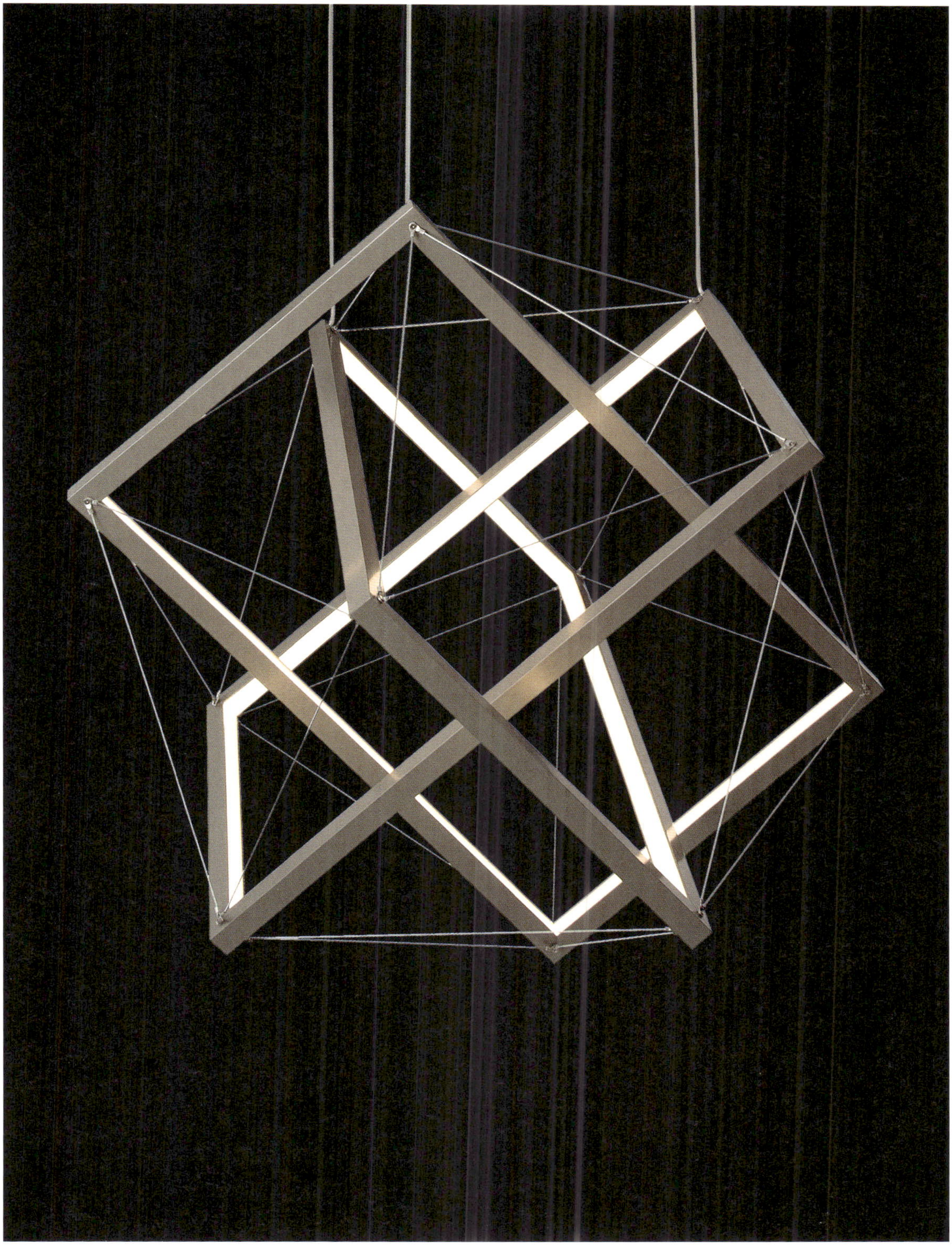

»The golden ratio is basic vocabulary, an absolute must-have.«

Fashion designer Christian Mau, Director of the BEST Sabel School of Design Berlin, in conversation with Katharina Schillinger & Florian Schütz

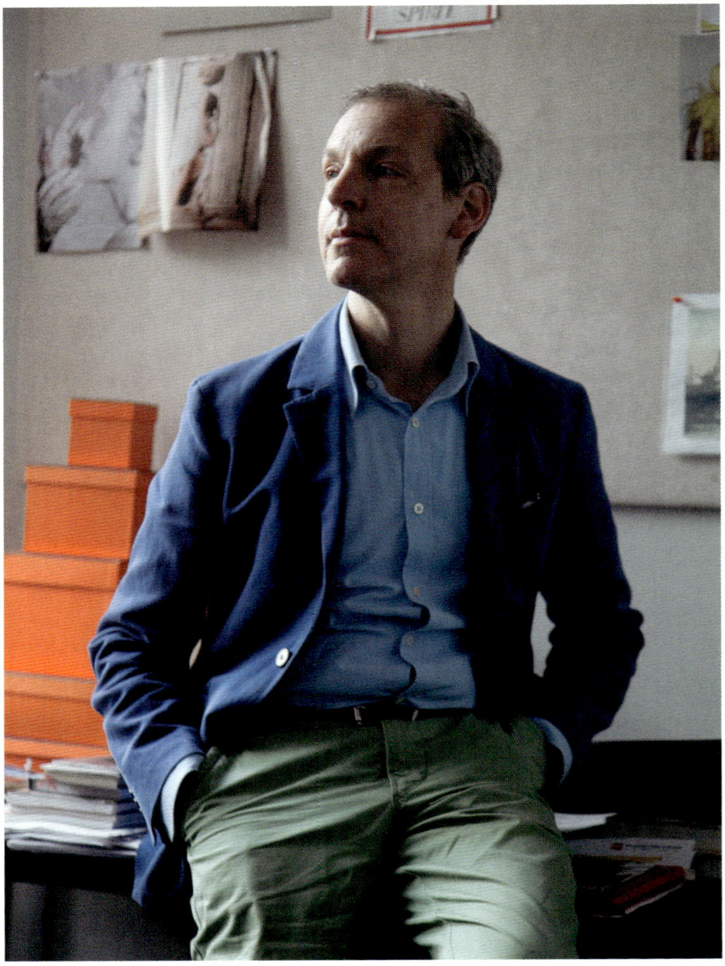

Christian Mau, BEST Sabel School of Design Berlin

Katharina Schillinger: Our exhibition presents many different facets of the phenomenon of the golden ratio. When and how did you come into contact with the golden ratio? Very early on! In my second year at university, we were each given pieces of material to tailor – a circle, a square and a golden rectangle. Though at the time, I didn't know that's what it was. (*laughs*) We were supposed to use these shapes to create wearable items of clothing, cutting them at little as possible. If you knew how to, you could use these basic shapes for any number of different proportional studies.

KS: Working on the basis of these reduced basic elements rather seems to contradict the credo ›form follows function‹. How would you work with the golden ratio if you had some kind of shape, say a pentagon, and had to create something from it?

Florian Schütz: No doubt something like that would be a challenge to an individual's creativity. Exactly, since even before you start worrying about the details, the objective is to develop ›radical‹ ideas about how our bodies move – not just their shape and structure, but also surfaces and the ideas of how to bring this together. When I've managed that and place it in today's context, this creates new ideas and expressions. But the first step is this creative, reduced decision-making process which, at the end of the day, ultimately has to be convincing.

KS: Was that your first encounter with the golden ratio, or had you come across it before? Through my parents and grandparents, I grew up with pictures. I was constantly thinking about how, for example, a picture, object or cupboard was designed, why the door was set at one point or another. So I was surrounded by well-proportioned objects and artefacts from early childhood on – even if I didn't know that's what they were at the time. But then my ›trained eye‹, as it were, always reacted quickly if something was not right – and I'd say, »The proportions aren't right!«

KS: So for you, does the golden ratio mean agreeable, harmonious proportions?

Absolutely! As designs, fashions and perspectives change, you need a kind of island to serve as a point of orientation – and that's the way it's been ever since I can remember. For fashion designers, proportions are everything. Colour, for instance, may also serve as a signal, but everything – our clothes, the things around us, the places we live in, as well as what we communicate, and even sounds themselves – they are all a question of proportion. »What form does something have?« is then the all-decisive question – which also has a philosophical dimension if something is to be properly understood.

KS: How important is the golden ratio in design, in teaching design, for communication to function between the sender and receiver? I believe the golden ratio is part of the basic vocabulary, an absolute must-have – and it's essential to know its structure and how it functions. In my view, this is a natural and necessary building block in teaching, no matter which area of design you're dealing with. Today, it's even more necessary than ever, since we are working at a time when design is in a process of change, and borders are shifting. We need to take these islands as a point of reference: »What was there in the past? What's happening now? What can we still do?« You can't always reinvent the wheel, sometimes you have to cleverly reassemble it.

At present, we're still monitoring how our senses react to the digital world which surrounds us everywhere. We are not yet able to draw any conclusions from this process, and we should indeed see it as a process. And as an older person, I try to give the golden ratio to my young students as part of their vocabulary, as a fact, but also as something you can then liberate yourself from again.

FS: Is the golden ratio always a safety net which can catch you if you're lost in other realms? Exactly – ›can catch you‹! And since there are different opinions, one has to accept that ›can‹. I put forward hypotheses that can be taken as something to work against, or may even be disliked. But if I have nothing at all, then I'm just going round in circles. But design always has a goal, and that goal is to solve some problem identified as such.

KS: In your view, is there an era or a figure you regard as especially relevant for the golden ratio and its dissemination? Le Corbusier, for example, gave a wonderful, universal architectural answer, away from applied industrial design, of what architecture can also be. And once you have seen his Modulor, you'll never forget it – it's something that sticks in your mind. Another example would be the photographer Erwin Blumenfeld, who used models in the 1930s to reconstruct classical harmony – and his work was one of my key experiences. Even today, I'm fascinated by the imaginative power that the Greeks utilised to convey and embody harmony, and the golden ratio as well, in such detail.

FS: Do you feel like that only about the golden ratio or also when it comes to other proportional relations that you also teach as must-haves? Do symmetries or a 2:3 ratio have the same significance? Looked at scientifically, there would be little difference in their importance, but the golden ratio simply is popular, benefitting from a kind of ›early PR‹. The events, theories or ideas which have always proved to be the most ›sustainable‹ in history are those which were cleverly marketed – for example, the French Wars of Religion, as they are known, which radically changed Europe in the late sixteenth century. There is so much well-researched material about them and they are so well-known because they were accompanied by the first major war PR campaign. That is the parallel to the golden ratio – good PR, an idea spreading like wildfire, a certain enigma with a high appeal factor. The golden ratio is louder, and may have just drowned out other theories which were certainly no worse.

KS: Well, the golden ratio may provide a point of orientation in design. But how important do you think it is to free yourself from it as well? Just as you engage with it, you also need to regain a distance, so you can feel how you experienced it, the impact it has on you – both on your mind and body. The golden ratio leaves traces, even if you don't see them. Our brains have stored it in the same way as good taste, as chocolate, and it can be reawaken after a long period of abstinence.

And that's why we can recognise the congruence of proportions and harmony intuitively as well as intellectually. Admittedly, this process is also dependent on a person's age, and it can change. Personally, I would argue that every person has this sense, and is confronted with it, or is searching for it. Perhaps you only find its forms, colours, and objects later in life – and then it becomes more difficult to give them up again.

KS: Do you have any examples of cases where the golden ratio does not work? Right, the golden ratio does not always function – for instance, in formal terms in the experimental design process. When you're experimenting, you have to free yourself, take a step back. Applying the golden ratio disturbs a) the experimental process and b) the search for something outside your normal thought processes. Or take a quite banal example: when you enter an empty apartment, in a split second you can imagine the furniture in relation to the room's proportions. In my case, I magnificently failed with this ›mental furnishing‹. Things that work in your mind taking the golden ratio as a measure do not then work in reality. So I also quickly stopped talking about it. (*laughs*)

KS: You mentioned that notions of beauty can be further developed individually or together with others. Would you generally say that the ideals of beauty change? Yes, they're constantly changing – in groups or individually, depending on your perspective and ideas.

FS: Can you also see this in your students? Certainly. As the process of teaching has changed, with more mentoring and coaching, you can also confront other topics more intensively. As a teacher, I can decide when I introduce the golden ratio building block – and that allows me to be more objective and open towards my students. In my view, that's a great advantage. You and your students reciprocally influence each other – and that is the ›honey‹ of teaching.

KS: It's not only the ideals which change – so too do the demands on design. Are economic considerations the priority? What determines design today?

FS: Love or money? Here, maybe I can first give you an example. I find the predetermined format of TV screens very problematic. Even my first TV – which, mind you, was already a flat screen – definitely seemed to be a disruptive element in my interior design. The device, the product, didn't want to fit in. I had to struggle with the TV's proportions, constantly wondering why they couldn't be different. Although this may sound like the problems of prosperity, it can nonetheless be quite painful. A technically determined shape is pre-given, and the ›analogue‹ customer has to respond to it. So love or money? In this case, I discovered that it's more about money.

KS: You talk about a disruptive element as an intruder in the private sphere. Do you find other proportions or other technical developments to be disruptive? This is an area where I can give you a whole list of things – starting with telephones. As soon as I see some artefact or hold it in my hand, I automatically check the proportions. The design of smartphones is too long, and I also take exception to the upright base stations for cordless landline phones, looking at me everywhere with their phallic shapes. Products we handle daily are poorly designed – telephones, tableware, cutlery. Admittedly, this is my own perception, but I'm happy to offer it as a point for argument and discussion with my students. If I didn't regularly exercise my mind in this kind of critical debate, I wouldn't know how others think and their perspective on things.

FS: The question you've raised is how many people consciously think about the design of the world around them. Many things are normed, and so often you have a very limited choice. Cars! Quite probably, on the spur of the moment I wouldn't find a single car outside that I'd personally consider attractive – even though there's a lot of money involved. But how do we take decisions? Rationally? No, we follow our instincts – that's what we do! I make a decision, and that's worth communicating, since through these decisions, we learn a lot about other people, and their priorities. I boldly assert that by critical analysing the golden ratio and communication, we are focusing on what we have in common – what is it that we need?

FS: Although deciding on other products, ones that are specially designed, usually also has a financial dimension. Don't we accept the ›standard‹ precisely because it is cheaper? Is good design expensive? The golden ratio is not expensive. It's available for everyone – and everyone can use it. But in our value system, investing time and intellectual energy in design also costs money. In comparison, in evolutionary terms, it also took nature a very long time to develop the necessary function of an aesthetic beauty which could be identified as such. If some ›standard‹ object is the result of a maturing process in design, then it can also be well-designed and trusted to be so. From that point of view, the economically ›expensive design investment‹ is transformed into an affordable raw material for a ›standard‹ product in the shops.

FS: Erik Spiekermann says: »Good design is not cheap. Cheap design is no good«. Where's the difference? Is the golden ratio a good tool to produce a valuable product? No, that's not how I see it. Instead, it is something that just relates to the process of critical thought. Thinking is work, and can be a painful process, and may lead to disappointments. The question is – what am I worth in my own eyes? Admittedly, this is a point where I'm too little a capitalist or manufacturer. I have to keep my cool when faced with some of those terrible products on sale. Stay friendly! (*laughs*) You have a responsibility for what you create, and fortunately, that view has also become more vocal – which doesn't mean that it's necessarily less creative. We ought to be more ready to experiment to shed these narrow constraints.

KS: But you would subscribe to the idea of the golden ratio as a classic and as timeless? Yes, I would. The current edition of *form* magazine is entitled »Designing Time« – and that fits in here with our discussion wonderfully. There are always times when we recall such classics as the golden ratio. Personally, I view the threshold between the early modern period and modernity as a decisive point. After that – also through the ›demystification of the world‹ – access to previously closed areas of fashion, colour theory, and aesthetics was opened up. The proverbial globalisation of that time opened people's eyes to ways of talking about and teaching design. Here, I can see parallels with today's discourse, since we are now again increasingly addressing issues in design, aesthetics and values.

KS: I would just like to recap what we've discussed. There are bad objects, whose design disturbs us; and there are good objects, which we like. But what about neutral things, those things we don't even perceive as designed? It is a compliment to these well-designed objects that they have insinuated themselves invisibly in our everyday life. That is probably indicative of the most outstanding quality. In my view, an unobtrusive discreet design is almost a luxury, a valuable quality. But here I'm not talking about uniformity. You can also buy something weird, something quirky, if it's the result of intensive thought – just as Erik Spiekermann says. Through haptic experiences, everyday things leave a feeling, a memory. The golden ratio is similarly connected to a feeling – and advocates of other ›cults of proportions‹ would say just the same.

FS: And there, we've reached everyday life. So what do you think of everyday standardisation – for example, international ISO paper sizes? I'd always imagined it rather differently. If I take a sheet of ISO A3 and trim it a little, then I feel at home. For me, there's another format, but I accept those standards that are generally valid. The ISO A4 sheet of paper is a social norm that is in circulation, and has far-reaching implications. But I'd actually prefer it to be different. In my view, everything I write or sketch on paper with other proportions looks better.

KS: So, are we locked into a vast pragmatic system? Communication is just simpler with normed standards. But some historical letters on display in the Museum for Communication have other formats. Their effect is completely different – and so is their message.

KS: Mr Mau, thank you – and many thanks for joining us for this discussion.

Of memes and snapshots
The golden ratio in the digital age

Melanie Lauer

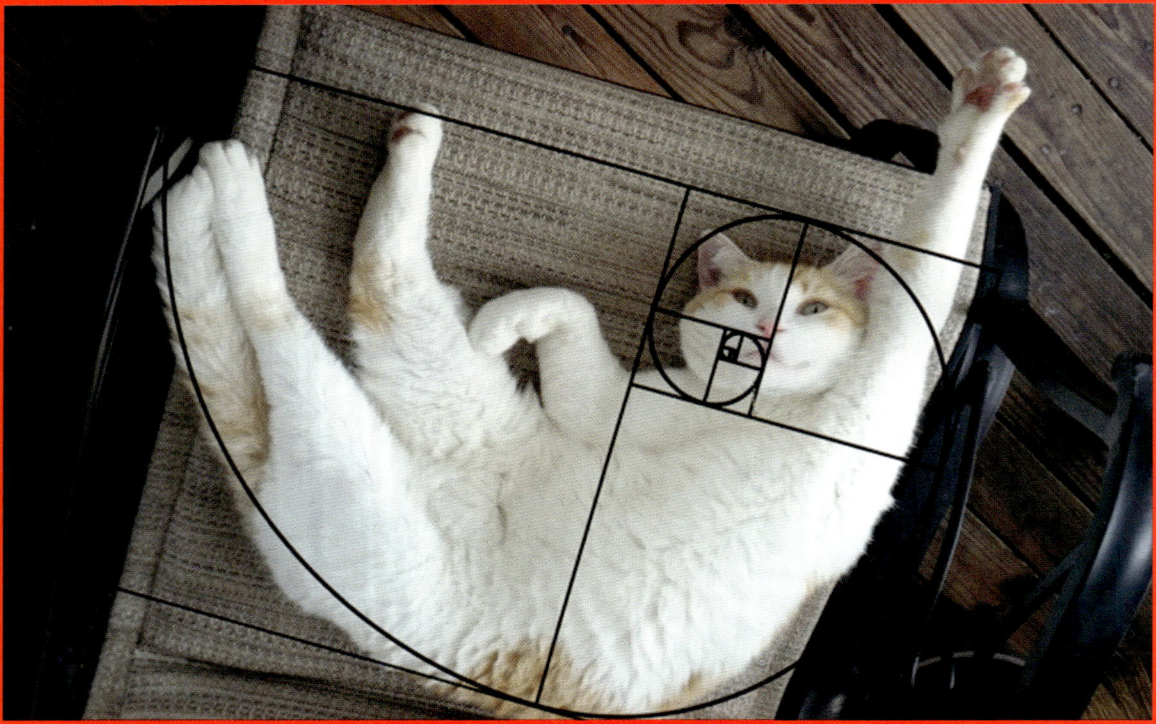

**Figure 1.
Golden
Cat-Spiral**

New Year 2016. Around midday on 1 January, the Manchester Evening News published a series of photos capturing scenes of the revellers in the city on New Year's Eve.[1] In itself, that was nothing out of the ordinary, yet less than two hours later the tweet of a BBC reporter led to one of the photos causing quite a furore. The image went viral, was commented on, shared and adapted. Soon it was even appearing in national and international media reports. The first internet meme of 2016 had been born.

Unexpected beauty on New Year's Eve

It all began when BBC reporter Roland Hughes saw the photo and tweeted »Like a beautiful painting«. In the wake of his comment, the cyber community took a closer look at the image by photographer Joel Goodman (Fig. 2, p. 204). The scene depicted is nothing that unusual on a New Year's Eve in Manchester. At a junction, some police officers are dealing with a man lying in the road. A young woman to the right is bending down towards them, while another man lying in the road nearby is reaching out for a beer bottle. Behind him, there is a queue of cars, and a small group of people on the pavement.

But something about the photo seemed to fascinate Hughes' followers. The Twitter community was swift to react, posting comments claiming that the photo had an artistic aesthetic comparable to Pablo Picasso or Joan Miró. Moreover, Dutch Twitter user @GroenMNG even noticed »interesting axi and regions« in the image, photoshopping onto the scene what is known as a golden spiral (Fig. 3, p. 205) – a particular logarithmic spiral recursively dividing each golden rectangle into a square and another, smaller golden rectangle.

The golden spiral evokes the golden ratio and the allegedly perfect proportions in the photo. This triggered a kind of chain reaction, and within a few hours of Roland Hughes' original Tweet, it had been retweeted over 20,000 times. Over the next couple of days, Joel Goodman's photo was uploaded in innumerable different versions.

How the meme conquered the internet

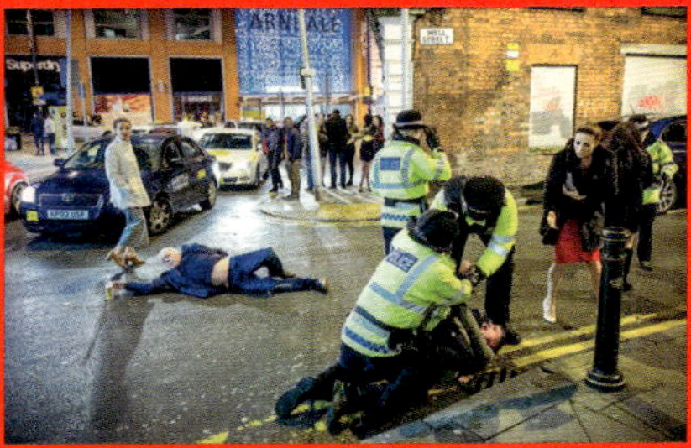

Figure 2. New Year's Eve 2015 in Manchester

Such massive global reactions in the form of internet memes, as they are known, have been part of popular online culture since the turn of the millennia, generally as part of Web 2.0 practices. In this participative and, in particular, collaborative approach to the World Wide Web »[...] users do not only consume content, but as prosumers actually generate content themselves«.[2] Journalist Dirk von Gehlen, who blogs for the Süddeutsche Zeitung newspaper on cyberculture, describes the phenomenon as follows:

»[Internet memes are] content which move through the Net very quickly, and seemingly by themselves. They are copied, altered, and disseminated so fast that it's no longer possible to say who actually invented them. As a result, [...] today you can constantly find new memes in the Net: images, sentences or sounds, set in a new context and passed on by viewers transformed into users.«[3]

The term ›meme‹, originally coined by Richard Dawkins, was already making an impact on the world of science and scholarship in the 1980s. In his book *The Selfish Gene* (1976), Dawkins transferred Darwin's theory of evolution to the world of thought, suggesting *meme* (a neologism from gene and mimeme) as a term to describe the way trends are self-propagating – which could apply equally to ideas, fashions or melodies. Dawkins explained that »just as genes propagate themselves in the gene pool by leaping from body to body via sperms or eggs, so memes propagate themselves in the meme pool by leaping from brain to brain via a process which, in the broad sense, can be called imitation.«[4] Since Dawkins first proposed this concept of the meme, it has been the subject of very diverse and differentiated debate in scientific and scholarly circles. However, according to Limor Shifman, Senior Lecturer in the Department of Communication and Journalism at the Hebrew University of Jerusalem, it is no coincidence that we encounter its digital equivalent – the internet meme – on an almost daily basis. She argues that »the Internet's unique features turned memes' diffusion into a ubiquitous and highly visible routine«, and notes that the three basic properties of memes as defined by Dawkins – longevity, fecundity and copying-fidelity – are intensely enhanced by the internet, and so offer memes a perfect breeding ground.[5]

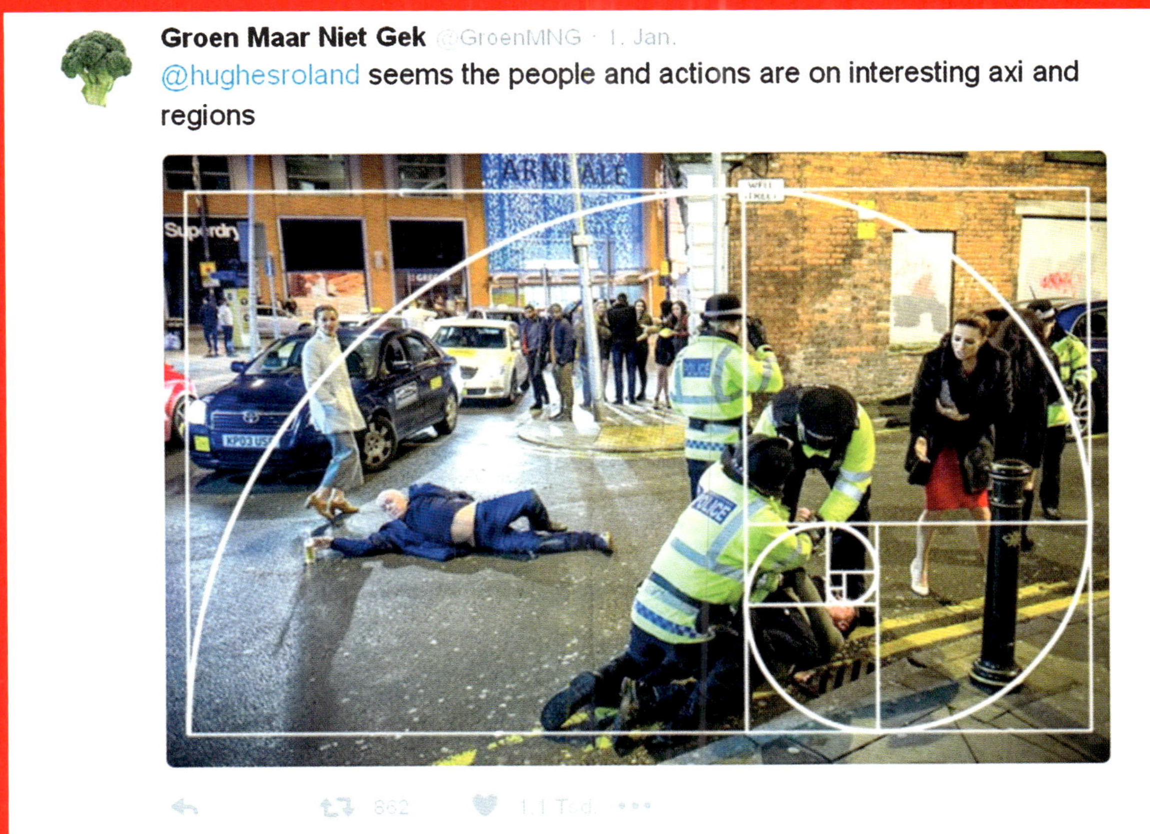

Figure 3. Twitter user @GroenMNG superimposed a golden spiral onto the photo

The phenomenon of the golden ratio and cyberculture

The idea of the golden ratio also seems to be particularly appealing to the cyber community. The American website knowyourmeme.com, for example, claims to have collected and documented internet memes and viral content since 2008. Every month, nearly ten million people visit the site, expanding and discussing the meme database which is managed as a wiki, a digital reference work whose content is modified collaboratively by its users. The website users have created a subpage dedicated to the golden ratio, as an especially popular topic, explaining that: »Online, the spiral (i.e. the golden spiral) is often overlaid on memetic images to demonstrate (either ironically or not) that these viral images are as pleasing as a classical composition«.[6] Hence, the Manchester photo providing the starting point for these considerations is far from being an individual case. On tumblr.com (the largest blogging platform worldwide with nearly 300 million blogs and over 130 billion blog posts), the search terms ›the golden ratio‹ or ›the golden spiral‹ produce a sheer infinite number of hits.

In addition, another interesting aspect is how those memes which seem to be especially popular often blend and merge. Online one can, for instance, find photoshopped montages of photos of Donald Trump, the US entrepreneur and Republican presidential candidate in 2016, whose controversial statements have made him the object of innumerable memes. Above all, with internet users across the world apparently being struck by Trump's ›noticeably proportional‹ hairstyle, there has been global dissemination of a photomontage with the golden spiral superimposed on his profile.

1 http://tinyurl.com/zbpxwql from 1 January 2016, accessed 15 April 2016.
2 https://de.wikipedia.org/wiki/Web_2.0, accessed, 18 August 2016 (English translation from the German).
3 Dirk von Gehlen, »Kontexte brechen: Wie Internet-Meme die Remixkultur demokratisieren – und jetzt auch das Fernsehen verändern«, in: Generation Remix. Zwischen Popkultur und Kunst, Djordjevic V./Dobusch L. (eds.), Berlin, iRights.Media 2014, p. 100.
4 Richard Dawkins, The Selfish Gene, Oxford University Press, 1976 (reissue 1999), p. 192.
5 Limor Shifman, Memes in Digital Culture, MIT Press, 2014, p. 17
6 http://knowyourmeme.com/memes/the-golden-ratio, accessed, 15 April 2016.
7 http://www.boredpanda.com/fibonacci-composition-cats-furbonacci/, accessed, 15 April 2016.
8 See Jürgen Fredel, »Dürer und der goldene Schnitt«, in: Die Beredsamkeit des Leibes, Salzburg und Wien, Barta Fliedl I./Geissmar C. (eds.), Residenz Verlag 1992, pp. 174–180.

The blog *Furbonacci Sequence Proves That Cats Are Purrfect* offers another vivid example of how several internet memes have been merged.[7] The blog is devoted to what it claims is the perfect proportions of cats. The title itself reads as a meme transcribed into writing, and contains several puns popular in the cyber community: *Furbonacci*, as a combination of *fur* and *Fibonacci* (since the ratio of any two successive numbers from the Fibonacci sequence will closely approximate to the golden ratio), and *purrfect*, a combination of ›purr‹ and ›perfect‹ are very common in content about cats. The blog, which lists over 50 images showing cats in various positions more or less approximating to the golden ratio (Fig. 1, p. 202), has been viewed over 125,000 times since it was uploaded.

Just like the photo from New Year's Eve in Manchester, these two examples can be taken as exemplary for the cyber community's playful and jokey approach to the golden ratio. The modern internet meme has nothing to do with the mystical readings or even spiritual idealisation of the golden ratio as pleasing, divine proportions or some kind of universal law, as so often found in the history of its reception.[8] Quite the contrary – in cyberculture, the concept (or the meme) of the golden ratio is viewed with considerable ironic distance, prompting internet users the world over to engage humorously with its supposedly aesthetic laws as well as, naturally, with the actual object of the meme – whether a controversial figure, an amusing comment or, indeed, a bizarre snapshot.

Beutelspacher, Albrecht/Petri, Bernhard: Der goldene Schnitt, Heidelberg/Berlin/Oxford, Spektrum Akademischer Verlag 1996.
Dawkins, Richard: The Selfish Gene, Oxford University Press, 1976 (reissue 1999).
Fredel, Jürgen: »Dürer und der goldene Schnitt«, in: Die Beredsamkeit des Leibes. Barta- Fliedl I./ Geissmar C. (eds.), Salzburg and Vienna, Residenz Verlag 1992, pp. 174–180.
Gehlen, Dirk von: »Kontexte brechen: Wie Internet-Meme die Remixkultur demokratisieren – und jetzt auch das Fernsehen verändern«, in: Generation Remix. Zwischen Popkultur und Kunst. Djordjevic V./Dobusch L. (eds.), Berlin, iRights.Media 2014, pp. 99–105.
Shifman, Limor: Memes in Digital Culture, MIT Press, 2014.

»The golden ratio could also be a ›golden road‹. At present, in the wake of a Darwinist anaesthesia altruism based on economics, dividing up in the spirit of a collectively valid aesthetics has even come to be a deplorable reflex. Yet sharing is one essential premise for the sustainability of human civilisation and culture. It may be presumptuous, but let me ask: Does the golden ratio offer the sharp division we need to redistribute global wealth – and wouldn't it be worth trying to take this principle on board in calculating our omnipresent algorithms? At any rate, one percent of the world's population owning more than the remaining 99 percent would then be a thing of the past.« Peter Zizka

Authors

Wolfgang Beinert

Typographer and graphic artist Wolfgang Beinert has worked for Leica, Chanel, Gmund Büttenpapier, Amnesty International and the Club of Rome. His works have been shown at the Frankfurt Book Fair, the Ginza Graphic Gallery Tokyo, the London Design Festival, and the Designweek Istanbul. He lives and teaches in Berlin where he also runs his studio, and addition, publishes the online typographer's encyclopaedia Typolexikon.de.

Albrecht Beutelspacher

Prof. Dr. Albrecht Beutelspacher studied mathematics and philosophy at the Eberhard Karls University, Tübingen. After working as a research associate at the Johannes Gutenberg-University, Mainz until 1982, he was appointed to a fixed-term professorship. Since 1988, he has held the chair in discrete mathematics and geometry at the Justus-Liebig-University, Gießen. He is also the founding director of the interactive museum Mathematikum Gießen, which opened in 2002.

Mark Braun

Mark Braun opened his own design studio in Berlin in 2006. His clients include such well-known companies as NOMOS Glashütte and THONET. His work has received numerous awards, including Design Plus, the Interior Innovation Award, and the German Design Award, and been shown in many galleries from the Saatchi Gallery to S. Bensimon, Libby Sellers, and Helm Rinderknecht. In addition to working as a designer, Mark Braun has held visiting professorships at, for example, HKD Burg Giebichenstein, Halle, and the ECAL Lausanne. He has been a full-time professor for industrial design at the HBK Saar since 2015.

Werner Busch

Prof. Dr. Werner Busch studied the history of art in Tübingen, Freiburg, Vienna and London, and presented his doctorate on William Hogarth in 1973. After working as a research associate in Bonn from 1974 to 1981, Werner Busch obtained his post-doctoral lecturer qualification in 1980. He held the chair in the history of art at the Free University of Berlin from 1988 to 2010. He is a member of the Berlin-Brandenburg Academy of Sciences and Humanities and ran the Collaborative Research Center 626 »Aesthetic Experience and the Dissolution of Artistic Limits« from 2003 to 2009.

Heike Büttner

In 1986, after studying architecture in Aachen and Berlin, Prof. Dipl.-Ing. Heike Büttner opened her own architectural office Prof. Heike Büttner (Büttner Neumann Braun, 1995–2001). Aside from her academic work at the Berlin University of the Arts, she has held teaching positions at the Munich University of Applied Sciences, the Zenobio Institute, Venice, and the Bauhaus University, Weimar (holding the chair in »Basics for designing« since 1995). Heike Büttner's work has been shown in many international exhibitions and awarded numerous prizes. She lives and works in New York, Berlin, Munich, and Weimar.

Claudia Erbar

Prof. Dr. Claudia Erbar studied chemistry at the University of Bonn. She was awarded her doctorate in 1982 for her work on organ sequence and carpel development in the flowers of some Magnoliidae at the Botanical Institute under Prof. Dr. Peter Leins. In 1983, she became a research associate in Heidelberg, and in 1993 went on to obtain her postdoctoral lecturer qualification. Claudia Erbar was appointed associate professor in 2000 at today's Department of Biodiversity and Plant Systematics at the Centre for Organismal Studies (COS), Heidelberg.

Katharina Fendius

Katharina Fendius studied culture, communication, and media in Berlin. Her fields of work include the cultural history of birth and death, the media history of the occult, body techniques and psychoanalysis. She has worked as a research associate at the Museum Foundation Post and Telecommunication since 2012 and is at present writing her doctoral thesis at the Humboldt University of Berlin.

Hartmut Fladt

After studying composition in Detmold, Prof. Dr. Hartmut Fladt wrote his doctor thesis on musicology in Berlin under Carl Dahlhaus. He has held professorial chairs at the Berlin University of the Arts and the University of Music and Performing Arts, Vienna. He is the author of over 90 publications on music from the thirteenth to the twenty-first century (including *The Music-Man: What we feel when we hear,* 2012, on popular music), and has also been a regular contributor on radio and TV. His compositions include chamber, orchestral, and electro-acoustic music as well as dramatic works, songs, and ›applied music‹.

Oliver Götze

Dr. Oliver Götze is the assistant director of the Museum for Communication Berlin. He studied the history of science and technology, mathematics, and modern history in Berlin and Venice, earning his doctorate for his research into the history of science and the art of the Italian Renaissance. He taught the history of media and technology at the Technical University of Berlin from 2006 to 2013. After working in various positions at the Museum for Communication Berlin from 2009 to 2013, he is now responsible for the Department of PR and Temporary Exhibitions.

Peter Hirschmiller

Peter Hirschmiller studied history and sociology in Freiburg and Konstanz, and went on to study military history and sociology at the University of Potsdam. He has worked for several history workshops and as a freelance provenance researcher. Peter Hirschmiller has been a research associate at the Museum for Communication Berlin since 2015.

Holger Höge

Prof. Dr. Holger Höge studied psychology at the Ruhr University Bochum. He completed his doctoral studies with a dissertation on the emotional foundation of aesthetic decisions and presented his thesis on the connection between consciousness and sound for his postdoctoral lecturer qualification. His research is focused on visual aesthetics, sound-processing, and the area of psychological museology. He was director of the Institute of Psychology at the University of Oldenburg from 2007 to 2010 and appointed honorary president of the International Association of Empirical Aesthetics in 2012.

Bärbel Högner

Dr. Bärbel Högner is an ethnologist and a photographer teaching at the Goethe University Frankfurt. Her main areas of interest include visual anthropology and urban ethnology. Her research work includes studies into the architecture and social implications of Le Corbusier's planned community at Chandigarh, India. Bärbel Högner lives in Berlin.

Mara-Lisa Kinne

Mara-Lisa Kinne studied the history of art, philosophy, and mathematics in Hamburg, Vienna, and Berlin. She is at present writing her doctoral thesis in the history of art at the Graduate Centre for the Study of Culture, Gießen. Her main research interests include the iconicity of mathematics, the genre of portraiture, as well as the formation of national identity in nineteenth-century art.

Lieselotte Kugler

Dr. Lieselotte Kugler studied history and art history in Bochum, and earned a doctorate in the history of art in 1983. In 1986 she was appointed as the founding director of the History Museum Saar, Saarbrücken. In 1999, she then became the general director of the German Museum of Technology, Berlin, and then director of the Museum of the Printing Arts, Leipzig. She has been the director of the Museum for Communication Berlin and head of the Museum Foundation Post and Telecommunication since 2007.

Marko Kuhn

After studying librarianship at the HTWK, Leipzig, Marko Kuhn was appointed as a librarian at the Francke Foundations, Halle (Saale) in 1999. He joined the Historical Museum, Leipzig in 2002, and has been head of the library since 2005. Marko Kuhn's main field of research is the history of Leipzig.

Franziska Land

Franziska Land studied textile and product design at the School of Art and Design Berlin Weißensee, the Royal Danish Academy, and the Burg Giebichenstein Kunsthochschule, Halle. During her studies she worked at the TextielLab of the TextielMuseum, Tilburg (NL). Franziska Land has been working for the Mark Braun Studio since 2015.

Melanie Lauer

Melanie Lauer studied history and media studies at the Ruhr University Bochum and the University of Strathclyde, Glasgow. After starting at the Museum for Communication Berlin as the press representative for online and social media activities, she is now the media representative of the Museum Foundation Post and Telecommunication and responsible for coordinating online strategies within the Museums for Communication in Berlin, Frankfurt, and Nuremberg.

Peter Leins

Prof. Dr. Peter Leins studied biology and chemistry in Tubingen, Freiburg, and Munich, where he also studied philosophy and earned his doctorate in botany. After obtaining his postdoctoral lecturer qualification, he held a professorial chair at the University of Bonn. From 1983 to 2002 he was director of the Institute for Systematic Botany and Plant Geography, Heidelberg, as well as director of the university's botanical garden and the Department for Biodiversity and Plant Systematics within the Centre for Organismal Studies (COS), Heidelberg. Although he officially retired in 2002, he is still actively engaged in research and teaching.

Daniel Lordick

Prof. Dr.-Ing. Daniel Lordick studied architecture at the TU Berlin and Carleton University, Ottawa, Canada. After obtaining his doctorate at the University of Karlsruhe in 2001, he worked as a research associate at the TU Dresden until 2010. He has also held a number of teaching positions, including visiting professor at the BTU Cottbus from 2005 to 2006 and, from 2010, associate professor at the TU Dresden. From 2011 to 2013, he was visiting professor at the TU Berlin in the department of architectural illustration and design. Daniel Lordick's main research focus includes geometrical modelling and visualisation.

Christian Mau

Christian Mau studied fashion design at the School of Art and Design Berlin Weißensee, the Berlin University of the Arts, and in Vienna. After working as an assistant designer for Wolfgang Joop in Hamburg and a design consultant for Giorgio Armani in Milan, Munich, and Hamburg, Christian Mau founded the design laboratory CMPdesign with Tina Siebert. He has shown his work in Berlin, Cologne, Graz, and Las Vegas, worked as a creative director and head of visual merchandising for Quartier 206 and as a creative consultant for Anne Maria-Jagdfeld Interior Design, Berlin. In 2013, he was appointed as director of the BEST Sabel Design School, Berlin.

Timo Niemeyer

Timo Niemeyer, who studied cultural anthropology and the history of art at the University of Zurich, is an art historian with a special focus on the European avant-garde. Since 2014, he has been managing an online portal for European art auctions in Berlin and Basel.

Katharina Schillinger

Katharina Schillinger studied the history of art as well as media and communication studies in Berlin. From 2008 to 2013, she worked in visual arts education for a variety of museums. She has been a research associate at the Museum for Communication Berlin since 2014, and curated the exhibition »Divine Golden Ingenious – The Golden Ratio as a Theory of Everything?« in 2016.

Florian Schütz

Florian Schütz studied history, religion and English at the universities of Bayreuth and Marburg. His main research focus is on the culture of remembrance, the history of the everyday life, the mentality and religion of the nineteenth century as well as colonial history. Florian Schütz is a research associate at the Museum for Communication Berlin.

Erik Spiekermann

Typesetter, information designer, typographer, and author Erik Spiekermann is professor emeritus at the Berlin University of the Arts and an honorary professor at the HfK Bremen. His many awards include an honorary doctorate from the Art Center College of Design, Pasadena, and an honorary award for his lifetime achievement presented by the ADC Germany. Erik Spiekermann's design studios in Berlin, Amsterdam, San Francisco, and Los Angeles are branded as Edenspiekermann.

Daniela Tisevski

Daniela Tisevski studied cultural anthropology and philosophy at the University of Basel. Since her apprenticeship as a manager for communication in 2000, she has worked in strategic marketing consulting, advising major corporations as well as small and medium-sized enterprises on design positioning.

Peter Zizka

Designer and artist Peter Zizka originally studied visual communication at the College for Design, Offenbach and is a founding member of the conceptual artist group MEMORY. In 1989, he also founded the design studio Heine/Lenz/Zizka with Achim Heine and Michael Lenz. His work has been shown at many international exhibitions and projects, and received numerous prizes. Peter Zizka is also a board member of the German Designer Club and the Art Directors Club Germany. He lives and works in Frankfurt (Main) and Berlin.

List of illustrations

pp. 72–83, **Fig. 1 to 7:** From the author's archive

pp. 86–91, **Fig. 1 to 5:** Architect's office: Büttner Neumann Braun GbR, Berlin; design: Prof. Dipl.-Ing. Heike Büttner; planning of execution: Prof. Dipl.-Ing. Heike Büttner, Dipl.-Ing. George Braun, Dipl.-Ing. Daniel Hänelt; construction management: Dipl.-Ing. George Braun, Dipl.-Ing. Peter Flucke; adviser of builder: Prof. Dipl.-Ing. Alois Peitz, Archbishopric Trier, Fig. pp. 86 & 90 from the author's archive, Fig. pp. 89 & 91 photographer Stefan Müller

p. 93, ***Instruments of Beauty: Divine Tools*, Olivia Lee, 2015**: The Primary Studio

p. 100, **Bird's-eye view of containers at the harbour in Hamburg:** Hafen Hamburg Marketing, photographer Michael Lindner

p. 104, **Portrait of Luca Pacioli, attributed to Jacopo de' Barbari, 1495:** Oil on wood, 99 x 120 cm, Museo e Gallerie Nazionali di Capodimonte, Naples

p. 111, ***Torlo*, Simone Stocker, 2005:** Bottoni

pp. 112–117, **Fig. 1 to 3:** From the authors' archive

p. 123, ***Les Trois* ceramic vessels, based on ancient amphorae for oil, wine and honey:** Studio Mark Braun, Granville Gallery Paris, 2012, photographer Andreas Schimanski

p. 124, **Trestle *m01*, a hybrid of the golden ratio and the Fibonacci sequence:** Studio Mark Braun, Thonet, 2014, photographer Magdalena Lepka

p. 128, **Flower model with fivefold radial symmetry, c. 1920:** Olbricht Collection, botany models, photographer Jana Ebert

p. 131, ***Dresden Madonna*, Bartholomäus Sarburgh, copy after Hans Holbein the Younger, *Virgin and Child with Burgomaster Jakob Meyer zum Hasen and his Family*, 1635–1637:** Dresden State Art Collections, Old Masters Painting Gallery

p. 132, ***Darmstadt Madonna*, Hans Holbein the Younger, *Virgin and Child with Burgomaster Jakob Meyer zum Hasen and his Family*, 1526/1528:** Würth Collection, Johanniterhalle, oil on softwood, 146.5 x 102 cm

p. 135, **»Circle of Beauty«, Zeising, 1854**: Zeising, Adolf, Neue Lehre von den Proportionen des menschlichen Körpers, aus einem bisher unerkannt gebliebenen die ganze Natur und Kunst durchdringenden morphologischen Grundgesetze, Leipzig, Rudolph Weigel 1854, p. 144.

p. 142, **Japanese funeral ceremony with tatami mats stacked against the wall, c. 1880:** Painting, ink, emulsion paint and gouache, 13.7 x 21.4 cm, Wellcome Library no. 565773i

p. 146, **Gustav Theodor Fechner (1801–1887), c. 1870:** Saxon Academy of Sciences, Leipzig, August Brasch, Photographisch artistische Anstalt, Leipzig

p. 153, **Golden Pineapple, Pols Potten, 2016:** Porcelain, glazed, photographer Peter Boesang

pp. 154–162, **Fig. 1 to 4**: From the author's archive

p. 164, **Y-fractal as three-dimensional analogue to the Pythagoras tree:** Dresden University of Technology, 2005, photographer Lutz Liebert

pp. 166/168, **Fig. 2 and 3:** From the author's archive

p. 171, ***Aquarius, Gemini*, Hanna Roeckle, 2014**: Hanna Roeckle, Aquarius, Gemini, 2014, car finish on SWISSCDF, each 24.5 x 19.5 x 19.5 cm, photographer Peter Hunkeler

p. 172, **Florence Cathedral, Samuel Manning, 1870:** Manning, Samuel, Italian pictures, drawn with pen and pencil, London, The Religious Tract Society 1870.

p. 186, **The Old Town Hall (*Altes Rathaus*), Leipzig, c. 1722:** Stadtgeschichtliches Museum Leipzig, Abbildung verschiedener öffentlicher und Privat-Gebäude zu Leipzig... Wie es theils war, theils noch ist. Ein Zeichenbuch für angehende Architekten. Mit 16 Kupfertafeln, Amsterdam, Buder & Sülpke 1811, sheet 1.

p. 193, **Pendant light *Golden Ratio*, Lutz Koebele-Lipp, 2016:** Mawa Design

p. 194, **Christian Mau, BEST Sabel School of Design Berlin**: Photographer Nithushan Sivalingam

p. 204, **New Year's Eve 2015 in Manchester:** Photographer Joel Goodman

p. 205, **Twitter user @GroenMNG superimposed a golden spiral onto the photo:** Twitter-Account https://twitter.com/groenmng

Imprint

Imprint

Divine Golden Ingenious
The Golden Ratio as a Theory of Everything?

A Museum Foundation Post and Telecommunication publication, published by Hirmer Publishers Munich to accompany the exhibition

Divine Golden Ingenious
The Golden Ratio as a Theory of Everything?

shown at the Museum for Communication Berlin
(09 September 2016 – 26 February 2017)

Also shown at
Museum for Communication Frankfurt (Main)
(26 March 2017 – 27 August 2017)

Publishing Editors
Lieselotte Kugler
Oliver Götze

With Contributions from
Wolfgang Beinert
Albrecht Beutelspacher
Mark Braun
Werner Busch
Heike Büttner
Claudia Erbar
Katharina Fendius
Hartmut Fladt
Oliver Götze
Peter Hirschmiller
Holger Höge
Bärbel Högner
Mara-Lisa Kinne
Lieselotte Kugler
Marko Kuhn
Franziska Land
Melanie Lauer
Peter Leins
Daniel Lordick
Christian Mau
Timo Niemeyer
Katharina Schillinger
Florian Schütz
Erik Spiekermann
Daniela Tisevski
Peter Zizka

Managing Editor
Florian Schütz

Project Management Publisher
Kerstin Ludolph

Proofreading
Lina Harder, Florian Schütz

English Editor
Christine House

Translations
Andrew Boreham
Brad Hagen (Beutelspacher; Lordick)
Patricia Newman (Leins/Erbar; Beinert)
Joe O'Donnell (Flact; Braun/Land)
Kate Sturge (Niemeyer/Tisevski)

Object Photography
Peter Boesang

Collections Museum Foundation Post and Telecommunication
Thomas Jabs

Library
Claudia Loest, Jan Neudeck

Design
Brousse & Ruddigkeit
Lea Brousse, Raban Ruddigkeit,
Anna Kusnier

Production
Katja Durchholz

Litography
Repromayer Medienproduktion, Reutlingen

Printing and Binding
Printer Trento S.r.l., Trento

Bibliographical data of the Deutsche Nationalbibliothek
The Deutsche Nationalbibliothek lists this publication in the Deutschen Nationalbibliografie; detailed bibliographic information is available on the Internet at www.dnb.de.

ISBN 978-3-7774-2692-1
Printed in Italy

www.hirmerpublishers.com

The book's layout is based on the golden ratio.
The proportion of 1:1.618 defines all spacings and layout rules. The typography is also set in this ratio.